Cruising

BRIAN KEANE has been a lifelong sailing since his days as a student, when he competed in an active Firefly dinghy fleet in Dun Laoghaire. Here he learned the rudiments of boat handling and racing tactics.

His work as an engineer, took him abroad for some years and on his return he became interested in the inland waterways. He acquired a motor cruiser and spent many happy times exploring the Shannon river and lakes. Marriage and the arrival of a family curtailed things for some time, but soon, the call of the sea was again answered. This time it led to a partnership in a Ruffian, which had and still has a thriving racing fleet in Dun Laoghaire. He then moved to the equally active Cruisers III fleet combining racing and cruising.

The enthusiasm for cruising has brought him to many ports of call throughout Ireland and across the Irish Sea, to England and Wales. He has also cruised in the Sporades in the Aegean Sea, the Caribbean, and recently completed an Atlantic cruise. He currently sails a Hunter Ranger. He is a member of the Royal Irish Yacht Club, Dublin Bay Sailing Club, and the Irish Sailing Association.

His interest in the sea is no doubt inherited from his maternal side. His grandfather who came from a long line of Boylans originally from Mornington in Co. Meath, obtained his square rig ticket in the 1880s. He was a master mariner in both sail and steam. A maternal uncle, Captain William Boylan, had a distinguished and often dangerous life at sea. He was awarded the O.B.E. for gallantry in seeing his crew survivors safely ashore in Derry, after his command, the *S.S. Yorktown,* was sunk by a U-boat 300 miles off Tory Island. That was September 1942 and he was in the now famous decoy convoy – the R.B.1.

CRUISING IRELAND

A GUIDE TO MARINAS AND MOORING BUOYS

BRIAN KEANE

Foreword by Mr Justice Robert Barr

The Alameda Press
Dublin

Published in 2000 by
The Alameda Press
P.O. Box 7902, Blackrock, Co. Dublin, Ireland

Copyright © 2000 Brian Keane

All rights reserved. No part of this book may be reproduced, stored in a retrieval system, or transmitted in any form or by any means, electronic, mechanical, photocopying, recording or otherwise, without the prior permission of the publisher. It may not be lent or resold in any cover or binding other than that in which it was published.

The publisher has made every effort to ensure that the information in this guide is accurate and as up to date as possible at the time of going to press. However, prices, telephone numbers and other details may change. It is emphasised that directions and maps are for general guidance and must not be used for the purposes of navigation. The publisher cannot accept responsibility for any consequences which arise from the use of this guide.

ISBN 0-9539269-0-7

Photos: Author
Rear cover photograph: The Fastnet Rock

Cover design and layout: graphics_it

Printed and bound by:
ColourBooks Ltd., Baldoyle, Dublin 13, Ireland

Visit our website for updated information - www.cruisingireland.com

This work is dedicated to the memory of my mother Katherine, from whom we inherited a respect for, and joy of, the printed word.

Contents

Foreword	9
Acknowledgements	10
Introduction	11
How to use this guide	13
Visitor Mooring Buoys	13
Ireland - Customs and Immigration Formalities	13
Emergencies	14
Pets	14
Money	14
Weather Forecasts	15
Eating out	15
Pubs	16
Gas	16
Personal Security	17
Golf (and other sports)	17
Church Services	17
Northern Ireland Telephone Numbers	17
Dublin	18
Dun Laoghaire	23
Wicklow	27
Arklow	29
Wexford	31
Kilmore Quay	36
Waterford	38
Helvic	41
Dungarvan	41
Ballycotton	43
Cork Harbour	44
East Ferry	44
Crosshaven	46
Kinsale	48
Courtmacsherry	51
Glandore	52
Baltimore	53
Sherkin Island	55
Schull	56
Crookhaven	57
Glengarriff	58
Adrigole	59
Lawrence Cove	59
Castletownbere	61
Dromquinna	62
Sneem	62
Derrynane	63
Portmagee	65
Knightstown	66

Kells	66
Dingle	66
Ventry	68
Smerwick	68
Fenit	69
Carrigaholt	71
Kilrush	72
Labasheeda	73
Foynes	73
Kilronan	74
Struthan	76
Maumeem	76
Kilkieran	76
Roundstone	77
Clifden	78
Leenane	80
Fahy Bay	80
Inisturk	80
Clare Island	81
Blacksod	82
Elly Bay	83
Ballyglass	83
Kilcummin	83
Teelin	85
Portnoo/Iniskeel	86
Aranmore	86
Downings	87
Portsalon	87
Rathmullan	89
Culdaff	90
Moville	91
Coleraine	91
Ballycastle	92
Carnlough	93
Carrickfergus	94
Bangor	95
Donaghadee	105
Portaferry/Strangford Lough	106
Ardglass	107
Carlingford	108
Malahide	109
Howth	113
Postscript	117
Further Information	118
Bibliography	119
Index	120

Foreword

Ninian Falkiner of immortal memory, who sailed the seas of western Europe from the Arctic to southern Portugal for many decades, was of opinion that the ultimate cruise is an odyssey around Ireland. It is only in recent years that I have taken his advice and two circumnavigations have now taught me how right he was. The secret is that Ireland, as an island should, reserves her natural beauty for mariners. Over a thousand miles of coast, including deep indentations such as Bantry Bay, Carlingford Lough and many more, provide a spectrum of contrasting delight and excitement. In my mind's eye I see the mist lifting in the morning sun and the cliffs of Slieve League appearing like a theatrical backdrop; the exhilaration of a strong following wind and a big sea through Inishtrahull Sound; the joy of a tranquil anchorage at Castletownsend when night is falling; the beauty of Derrynane in repose; the welcome anticipation of Lawrence Cove; homecoming at Lough Swilly with Knockalla beckoning like the minarets at Mecca to a devout Arabian; the Tuskar Rock in angry mood to windward. These are the stuff of ultimate cruising.

But there are other crucial ingredients of the Round Ireland experience - the places and the people you meet and the historic fabric which binds it all together. Brian Keane has brought these ingredients alive with great skill. *Cruising Ireland* is a real delight. It is full of practical information, invaluable for anyone exploring the Irish coast; but it also provides the discerning sailor with an artist's impression in historical terms about how Ireland and the Irish have evolved in the last Millennium and more. The end result is a treasure which will add much to the success and enjoyment of many a cruise in Irish waters.

Robert Barr
August, 2000

Acknowledgements

Compiling a guide was never going to be easy, with a great deal of information needing to be gathered, checked and recorded. This task could never have been completed without the support of my family and friends who happily gave time from their busy schedules. My wife Dervila was, as ever, very supportive as she has always been, whilst waiting for me and my crew to land safely.

My son Paul (when not otherwise occupied racing in Cork and Cowes!) and Janet King gave generously of their creative talents in designing the cover and layout, and in pulling the text, maps and photographs together. My daughter Louise gave immensely valuable advice based on her own publishing experiences in London. But the list does not stop there; sons Peter and John were willing to proof read and check. Alan McGettigan gave so generously of his time as we scoured the coasts of several counties for information on locations. Judge Robert Barr was always encouraging and gave invaluable advice, and graced the book with a foreword. Séamas ÓBuachalla kept me straight on the historical inclusions, pointing out quite a few facts that needed clarification or adjustment. Margaret and Jim Linnane welcomed me ashore in West Cork and were always ready with hospitality and useful information.

But it does not end there; so to Karen Hope of the Northern Ireland Tourist Board, Liam Campbell and Brian Geraghty of Bord Fáilte, Ann Wilkinson of The Marine Institute, Bernard Gallagher of B.J. Marine Ltd. and Alan Usher of Windmill Leisure and Marine Ltd, my thanks for all their help and encouragement.

Introduction

The coast of Ireland offers wonderful opportunities for the cruising yachtsman. Circumnavigation of the island of Ireland entails a journey of some 700 miles. But this distance can stretch to well over a 1000 miles if visits are made to the many loughs and estuaries which indent the marvellous coastline. There are numerous islands, especially off the west coast that provide a change from mainland living and often give great insights into Ireland's social and historical past. The mountains, coastal cliffs, inlets and islands, conspire to deliver possibly some of the most breathtaking scenery in western Europe. Given the right conditions, with the sun, perhaps after a rain shower, slanting on hills, shading valleys, etching the lines of cliffs, the resultant colours can be magical. Couple that with a friendly greeting ashore, good places to eat and drink, sites with compelling histories to visit, things to do; it all adds up to a journey very worthwhile.

One thing Ireland is not short of is history. Although described by one writer as an 'offshore island off an offshore island', it has been at the forefront of so many endeavours. From early Christian monasticism, which it helped spread across Europe, to literature, science and latterly sport and entertainment, Ireland, with a population today of only four and a half million people, has made quite an impact. From its original Celtic origins, Irish culture has been significantly influenced by various invasions; Viking, Norman, French, Spanish and English. The Viking influences are ever-present in the place names of the east coast; Howth, Wexford, Arklow. And despite wars, internal discord and famine, the Irish have maintained a sense of pride, their spirit and of course, their humour. It was at times by putting a witty or wry slant on things that the Irish maintained their composure. As you travel around the coastline, drink and eat in the pubs and restaurants, seek out the local history and exchange banter with the locals, this will become clear.

For the overseas visitor, Ireland is that bit nearer perhaps than is generally imagined. From The Mull of Kintyre in Scotland to Fair Head in the north east 'corner' of Ireland is only 20 miles. From Holyhead in Wales to Dun Laoghaire is 55 miles. Travelling from Land's End in England and from Brest in France to Cork covers 200 and 300 miles respectively.

The best time to travel is probably mid-June through July to mid-August; the optimum chances of good weather usually occurring around this time, and daylight lasting for up to 18 hours. It must be noted though that even at the best of times the Irish summer weather can be quite unsettled. Gales can be expected in summer, perhaps on two or three occasions even during this period. However, the pleasant weather can stretch right into September. The Irish Meteorological Service (Met Eireann) provide detailed weather forecasts and give the prudent skipper ample time to plan a safe passage, and find a sheltered berth before the weather deteriorates. This provides an opportunity to go ashore to visit and explore the local scene. Then, before too long the weather will settle and it will be time to set sail again. Always build enough time into your journey to sit out bad weather. Many a cruise has been spoiled by crews having to reach a destination by a certain time and bearing the brunt of poor weather. The old adage is so true – 'gentlemen never beat to windward whilst cruising'. No, those wily souls have time to enjoy themselves whilst waiting for a favourable wind. Sailing in up to a force five in Irish waters will give a reasonably comfortable passage. Consult a tidal atlas as part of your passage planning; having a favourable tide under you, especially on the east coast, will add considerably to your over the ground speed. The west or Atlantic coast of Ireland should be treated with care. Even on a calm day, a heavy swell can prevail.

One note of caution; when sailing in Irish coastal waters, skippers should be aware of drift nets. These are often placed off headlands and can stretch considerable distances perhaps across the boat's track. Keep an eye out, the ends are marked by buoys, with a fishing boat often stationed at one end, and steer around them. Also watch out for lobster pots marked by a single buoy of any shape or colour.

The idea for this guide had been in my head for some time but was given impetus, by two events. The first was the placing of visitor mooring buoys by Bord Fáilte – The Irish Tourist Board and The Marine Institute, in 39 locations around the coast of Ireland. The second and connected event was the publication in 1999 of a promotional leaflet in English, French and German, encouraging sailors to visit Irish waters and use the welcoming necklace of marinas, pontoons and buoys, right around the coast. The next and logical step was to provide visitors with a guide to these locations to help them get maximum enjoyment and satisfaction from their stay.

There are many ways for the cruising yachtsman to ascertain how to make a safe landfall in Irish waters. These include reference to the many excellent publications including the two pilots for north/east and south/west coasts published by the Irish Cruising Club, Macmillan's Almanac, and of course charts. It must be strongly emphasised that this guide is in no way a substitute, and should not be used for navigation. But these publications generally stop once the mooring buoy is secured, the anchor is holding or you are safely tucked into a marina. This guide takes it up at that point. Finally, talking to those who have already 'been there' is important, and when in a location and planning the next leg of the cruise nothing surpasses the knowledge of locally based sailors and fishermen.

How to use this guide

The Guide is laid out as an imaginary clockwise circumnavigation of Ireland, starting at Dublin. After Dublin, brief directions to Dun Laoghaire are provided followed by Wicklow, Arklow and so on. However, as a result of this construction Howth is the final entry and I hope my good friends there will understand why. In any case there is an index which provides a quick location to any destination.

A total of 70 locations (see map page 33) have been chosen for inclusion. This in no way precludes the reader from choosing any of the safe anchorages that are described in the pilots and charts, and not included here. But this guide will help the crew with, as befalls us all, limited time, to get the most out of the time available. Instructions are given for entering marinas, mooring pick-up, and coming alongside pontoons. For many locations a map showing main streets and facilities is provided. As already stated, it is recommended that this book is used in conjunction with the appropriate charts which are referenced for each location and the pilots which are listed at the back of the guide.

There is a brief description of each location, its origins and history. Sites of particular historical interest, and how to visit them are described. Restaurants and pubs within easy walking distance are listed (Dublin venues may require a taxi). In addition, all the facilities that visiting yachtsmen seek are also listed; these include telephones, banks and the Post Office (when within easy reach), fuel and water supplies, showers and toilets, shops, supermarkets and launderettes. And of course the necessary sources of hardware and servicing are identified; chandlers, riggers, sail makers, and engine, hull and electrical repairs.

Visitor Mooring Buoys

Visitor mooring buoys are laid and maintained by the local council of the county in which they are located. For example, the buoys in Ballycotton are maintained by Cork County Council. The European Union funded 75% of the half million pound cost of providing the buoys. They are distinguishable from other buoys by being cylindrical with a flat top and are yellow in colour. They can hold craft up to 15 tons in all weather and tidal conditions. There is a ring on top of the buoy through which a mooring warp can be led. There is a charge of approx £5 per day. Some have not been laid in the most convenient of positions, but they will at least give secure holding, until you decide on an alternative.

Ireland – Customs and Immigration Formalities

This guide covers all 32 counties of Ireland, 26 constitute the Republic of Ireland, and six the province of Northern Ireland, which administratively is part of the United Kingdom. The Republic of Ireland and the United Kingdom are members of the European Union, and citizens of these and other European Union countries are not required to go through immigration facilities. However, it is advisable to carry a passport or other form of identification, and proof of vessel ownership. Vessels arriving from countries outside the EU are required to obtain customs clearance.

Emergencies

Distress signals should be transmitted on Channel 16. Emergency situations in the Republic of Ireland are monitored by the Irish Coast Guard. They co-ordinate responses by the Coast Guard helicopter service, Irish Naval Service vessels, Irish Air Corps, Royal National Lifeboat Institution, An Garda Síochána (Police) and Coast and Cliff Rescue Service. They mutually co-ordinate with Belfast Coastguard (Maritime Rescue Sub-Centre) covering Northern Ireland, keeping watch on Channel 16, and liaising with the Royal Air Force, Royal Navy, Royal National Lifeboat Institution and Police. Both jurisdictions liaise with the British mainland.

Mariners can be assured that a comprehensive rescue service is in place ready to respond immediately to all emergencies.

Pets

The island of Ireland is rabies free. Pets from outside Ireland and the United Kingdom may not be brought ashore. This law is vigorously enforced and there are heavy penalties, including confiscation of animals for breaches.

Money

The currency of the Republic of Ireland is the Irish Pound, which is a Euro currency. One Euro equals .787564 Irish Pounds. Euros and Cents come into circulation on January 1 2002. For six weeks from that date the Euro and Irish pound co-circulate.

The Irish Pound is available in denominations of £50, £20, £10 and £5 notes. Coins are £1, then 50p, 20p 10p, 5p, 2p and one penny.

The currency in the six counties of Northern Ireland is the pound Sterling. It is not a Euro Currency. There are £50, £20, £10 and £5 notes, and £2, £1, 50p, 20p, 10p, 5p, 2p, and one penny coins. Some banks in Northern Ireland, issue their own notes. These should be changed before leaving; they are not generally accepted in mainland UK. Some shops, bars and restaurants in border areas will accept UK currency and vice versa, but this cannot always be relied on.

The Euro will be available in notes of 500, 200, 100, 50, 20, 10 and 5 denominations – and two euro, one euro, and 50c, 20c, 10c, 5c, 2c and 1 cent coins.

Weather Forecasts

Weather forecasts, gale and swell warnings for the waters around Ireland are provided by Met Eireann, the Irish Meteorological Service, as follows:

National Radio:

RTE 1, live on the following frequencies:

MW - 567 kHz (this is a powerful transmitter and can be picked up 100 miles or more off the Irish coast)

FM - 88.2, 88.5, 88.8, 89.1, 89.2, 90, 95.2 MHz

The broadcast times are: 0602, 1253, 1855 (Saturday & Sunday), 1902 (Monday-Friday) and 2355. Gale warnings are broadcast by RTE 1 at the first programme break after receipt and with news bulletins.

Coastal Radio:

Irish Marine Coastal Radio Stations rebroadcast the Sea Area Forecast every three hours beginning at 0103 UTC, after an initial announcement on Channel 16.

Gale warnings are broadcast as soon as they are issued, on the next hour and thereafter at 0033, 0633, 1233 and 1833 LOCAL TIME.

Stations and channels are: Malin Head 23, Glen Head 24, Belmullet 83, Clifden 26, Shannon 28, Valentia 24, Bantry 23, Cork 26, Mine Head 83, Rosslare 23, Wicklow Head 87 and Dublin 83.

Tel/Fax:

The Sea Area Forecast, together with any relevant gale or swell warning, can also be obtained through Weatherdial tel: 1550 123 855 and fax: 1570 131 838 (Code 0010 for full Marine Products list).

The BBC broadcasts a comprehensive forecast for all the sea areas around Great Britain and Ireland on Radio 4, 198kHz (1515M), at 0535, 1201, 1754 and 0048.

Eating out

That most important of activities after a hard days sailing – an enjoyable meal in good company and pleasant surroundings, is well catered for. Ireland offers a fine selection of restaurants, offering local and succulent seafood and meat, many with ethnic themes. Eateries may be divided into four categories.

- **Hamburger outlets, 'fish and chippers', cafes;** these offer budget meals, many for takeaway, from a couple of pounds up and generally do not serve alcohol.

- **Pubs serving food;** Most Irish pubs now serve food, many all through the day. Food can range from sandwiches and snacks to full meals, and most serve food of good quality. Sandwiches cost about £1.50, main courses from £5 to £10.

- **Medium-priced restaurants;** The majority of Irish restaurants fall into this category. Starters are in the range £2.50 to £5, main courses £10 to £15 and desserts £3 to £5. A three-course meal with a half bottle of house wine will generally cost £20 to £25 per head. All restaurants have licences to serve wine, some have full licences and serve beer and spirits in addition.

- **Expensive restaurants;** Where restaurants are known to be in this category it is mentioned in the guide. Here prices can range from £30 to £50 per head, perhaps higher for a three-course dinner and half bottle of house wine.

Most restaurants add a service charge of 10% on the bill. If there is no service charge, the usual tip for good food and service is around 10%.

Pubs

The Irish pub is a somewhat unique institution – a place to quench thirst, get into conversation with the locals, exchange experiences (nautical and otherwise), or just sit and contemplate your lot in life. Stout, lager and beer are the popular tipples. In the Republic, a pint (half litre approx) of stout ranges from £2.10 to about £2.40. For beer and lager add about 30 pence per pint. If you order a half pint of stout or beer ask for 'a glass' of your preference. Whiskey £2 to £2.50 for a half measure (colloquially - 'a small one'). But a large whiskey or any spirit is a double small and is also called 'a glass', just to complicate things! A gin and tonic ranges from £2.50 to £3.50, a glass of wine at about £1.50 to £2.

In Northern Ireland prices in sterling are about the same, but the spirit measure is about two thirds of that south of the border.

Gas

This is generally available in the larger locations but it is advisable to keep some in reserve in case you cannot locate a replacement quickly. Locals will usually be able to advise on availability.

Personal Security

Personal security is not a problem in the majority of locations described. The centre of Dublin, if treated in a commonsense way, should not be a worry. But there have been incidences of mugging of tourists. So some tips as given by An Garda Síochána (police in the Republic of Ireland): avoid poorly lit or streets with few people about; do not carry large sums of cash; do not display expensive jewellery; the obvious comments apply to ladies handbags – keep them very close to the person.

The situation in Northern Ireland has been transformed with the cease-fire and the recent positive political settlement and the roads and streets have thankfully recovered the appearance of what one finds elsewhere in western Europe.

Golf (and other sports)

Ireland has very many golf courses, so you are never far from one. It is safe to say that an Irish town of any reasonable size boasts at least one course. In addition many hotels have courses attached. Courses range from a simple nine holes to championship links. Names like Portmarnock, Royal Portrush, Ballybunion and Killarney are world famous. So yachtsmen who like to hit the white ball will never be far from a course, in the majority of locations covered in the guide. Local people will gladly advise. If you don't pack your golf bag in the bilge, most courses have clubs for hire.

Ireland is a very sports active country, and facilities for other popular sports such as tennis, squash, swimming are usually available in the larger towns.

Church Services

Locals will supply information on church services. In the Republic of Ireland Roman Catholicism is the majority religion, and you are never far from a Catholic Church. Other denominations are somewhat less well represented especially outside the areas of large population. In Northern Ireland in the locations covered, churches of the Protestant and Roman Catholic denominations will be found.

Northern Ireland Telephone Numbers

All local Northern Ireland telephone numbers have been increased to eight digits. The old numbers can still be used up until April 2001. All codes will commence with 028 with two or three extra digits added to the local number(*LN*) to bring it up to eight. Examples – 012657 becomes 028 207*LN*; 01960 becomes 028 93*LN*; 01247 becomes 028 91*LN*; 01396 becomes 028 44*LN*; 012477 becomes 028 427*LN*; 01232 becomes 028 90*LN*. An example – existing number 01247 467699 becomes 028 9146 7699. The access code from the Republic of Ireland has been changed from 080 to 048.

Dublin

Latitude: 53°20'.50N; Longitude: 06°15'W;
Charts: Admiralty 1447 1468; Imray C61 C62

The entrance to the port of Dublin, where the river Liffey meets the sea, is at the western end of Dublin Bay. This sweeping and panoramic bay is framed by the Dublin and Wicklow mountains to the south and the Hill of Howth to the north. The bay, mountains and sky provide a handsome backdrop to the buildings, the spires and landmarks. Approaching Dublin Bay, the Kish Lighthouse stands guard about ten miles east of the entrance to the port. It marks the northern end of the Kish Bank. It was prefabricated in nearby Dun Laoghaire in the 1960s, and successfully towed out and lowered onto its prepared foundation. The North and South Burford buoys mark the ends of another bank – the Burford.

The entrance to the port is clearly buoyed, and once abeam the red tower of the Poolbeg lighthouse, you are in the walled channel. To the east is the Bull Wall and to the north not surprisingly, the North Wall. The original surveys which led to these walls being built to provide protection, guide the flow of the river Liffey and avoid silting, were carried out by none other than Captain Bligh, the notorious master of *HMS Bounty*. Dublin is a busy port, so keep a watch for ships, pilot boats and high speed catamarans. The East Link Lifting Toll Bridge will soon be clearly visible, with the Poolbeg Yacht Club and moorings to the left of the bridge. The bridge can be opened at 1100, 1500 and 2100 hrs. Visiting boats should contact the Berthing Master on Channel 12 and stand by for instructions. Dublin Docklands Development Authority have installed a mooring pontoon alongside their headquarters on the North Wall Quay. Once through the bridge the pontoon is clearly visible on the north (right hand) side of the river. Having tied up, you will be within ten minutes walk of the capital's city centre. Electricity and water are available. There is a shower and toilet block. Initial accommodation is for 25 boats. Rates are £1.50 per metre per day. Booking is not necessary but the Authority telephone number is 01 8183300.

What follows cannot hope to provide a comprehensive guide to Dublin and for a more complete coverage the reader is referred to the many excellent publications available, such as *Fodor, Time Out, Lonely Planet, Dorling Kindersley/Eyewitness*. The budget minded could try *Dublin on a Shoestring*. However, it will hopefully give an insight, as to what awaits the visitor in this great and historic city.

Dublin derives its name from the 'Dubh Linn', or 'Black Pool', formed in the Liffey, where the Norse settlers of the late ninth century built their settlement. By the end of that century they had established a flourishing trading post and fort, and proceeded to extend southwards. Many Norse remains, indeed whole streets, have been uncovered during recent building works and the National

Museum houses a comprehensive collection of household and other artefacts from that period. Recent excavation has uncovered what may be Saxon remains beneath Viking habitations.

Norse power started to decline in the tenth century, and was extinguished with their defeat at the Battle of Clontarf in 1014 by the native Irish, under Brian Boru. The Irish reign was in turn to be overthrown, following the invasion by the Anglo-Normans under Strongbow. Despite their valient efforts, the Irish failed to defeat the new invaders. Henry II of England subsequently granted Dublin a Royal Charter, and the development of the city began with the construction of such buildings as Christ Church and St. Patrick's Cathedrals, which since then, have been in constant use as places of worship.

The English hold remained firm with various attempts to dislodge them failing. In the early seventeenth century the development of Dublin as a major city and port began. This development peaked in the eighteenth century with the establishment of the Wide Streets Commission. From this came the splendid Georgian squares and streets such as Sackville Street (renamed O'Connell) and Fitzwilliam Square. Plaster ceilings and marble fireplaces were decorated by Italian craftsmen. A series of graceful public buildings was constructed, unchanged to this day and there to be admired – The Irish Parliament, now the Bank of Ireland, the Customs House, Trinity College, the Royal Hospital, the Four Courts, Leinster House (now housing the Oireachtas or Irish Parliament). Dublin was to be the first European city to be graced with such splendid classical buildings, and was dubbed the second city of the empire after London. It was chosen as the venue for the first performance of George Frederick Handel's oratorio – *The Messiah*. The Phoenix Park still the largest enclosed park in Europe was laid out. Its boundary wall is seven miles in length.

In 1801 with the passing of the Act of Union, the Irish Parliament was dissolved, the city went into a period of decay, exacerbated by punitive tariffs on exports. Ireland was absorbed into the United Kingdom. The splendid Georgian squares especially on the north side of the Liffey were deserted by the peers, MPs and administrators, and became slum dwellings. This situation would not be rectified until well into the last century. Life in these tenements was immortalised in the plays of Sean O'Casey.

The desire of the Irish for self government and independence was to surface in every century, in 1641, 1798 and throughout the nineteenth century. A Home Rule Bill to provide self government was passed in 1912, despite objections by northern unionists. However its implementation was delayed, because of the outbreak in 1914 of the First World War. But at Easter 1916, a group of Republicans led by Patrick H. Pearse, and a socialist citizen army led by James Connolly, occupied the main public buildings in Dublin, and declared Ireland to be an independent republic. After a week of fighting, hopelessly outnumbered by British forces, the insurrectionists surrendered. Much of the centre of Dublin was in ruins, reduced to rubble by the guns of *HMS Helga*, which was moored in the Liffey. The British reaction was to summarily try and execute 16 leading republicans by firing squad in Kilmainham Jail. Many were poets and intellectuals. Pearse himself, an educationalist and creative writer, had a fatalistic streak, knowing he would die, but remained convinced that it was only though spilling blood that Irish republican aspirations would be fulfilled.

Public reaction which had been hostile, suddenly became sympathetic. Amongst those executed, was James Connolly, the leading socialist thinker who although badly wounded was nonetheless shot. There was to be no peace, as Sinn Fein - 'ourselves alone', swept to power in the post-war election of 1918. A guerrilla war ensued, largely masterminded by Michael Collins. In 1921 the two exhausted sides agreed to a treaty, which would give dominion status, which Canada enjoyed, to 26 of the 32 counties of Ireland. Six counties with a unionist majority would remain part of the United Kingdom. Dáil Eireann, the Irish parliament, voted by a narrow majority to accept the treaty. But the opposition walked out and a civil war ensued. The country's infrastructure such as it was, suffered severely. It was to be a costly and devastating episode.

Peace gradually returned and the fledgling state started the job of reconstruction and development. Progress was slow. A policy encouraging local industry, protected by high tariffs was pursued, which would reduce the traditional dependence on agriculture. For the period of World War II, Ireland remained neutral, although many individual Irishmen enlisted in the Allied forces. It was not until the late 1950s that a more open economic policy was implemented. A major drive to increase skill and educational levels was implemented, foreign investment was actively pursued. In 1973, Ireland joined the then EEC. Farming was transformed as Irish produce now had access to a wider market, with better prices guaranteed. This contrasted with previous dependence on the British market where prices were depressed because of imports from commonwealth countries. The results of that long process are apparent today with the country dubbed 'The Celtic Tiger', and enjoying a high standard of living.

Dublin has long been 'fertile ground' where the arts flourished. From Jonathan Swift who wrote *Gulliver's Travels* while he was Dean of St. Patrick's, the city has produced many writers and playwrights – Richard Brinsley Sheridan, George Bernard Shaw, Oscar Wilde, Sean O'Casey, James Joyce, Brendan Behan, Samuel Beckett, W. B. Yeats; not forgetting Brian O'Nolan alias 'Flann O'Brien' alias 'Myles na Gopaleen'. The National Theatre (The Abbey) founded in 1904, stages the best in Irish and imported drama. There is the Gate Theatre founded by Micheal MacLíammóir and Hilton Edwards in the 1920s, to stage works then not performed in Ireland, such as the plays of Ibsen. It still flourishes. On its boards famous names such as Orson Welles, James Mason and Harold Pinter had their first breaks. There are many other theatres, and the National Concert Hall stages regular concerts and recitals.

The best way to bring this history of the city to life is to take a stroll around the streets of Dublin. The two major art galleries are the National, and the Municipal which houses the Hugh Lane collection. Lane tragically perished in the sinking of the *Lusitania* off the Old Head of Kinsale in 1915. Trinity College, founded in 1592, houses the world-renowned fifth century Book of Kells, one of the finest examples of illuminated gospels extant. The National Museum has a comprehensive range of artefacts, reflecting the people and their history. For shopping try Grafton Street, somewhat upmarket, or Henry Street a bit more down to earth. The pubs of Dublin are numerous and famous for their banter. The Temple Bar area is Dublin's answer to the Left Bank and its narrow cobblestoned streets house a range of pubs, eateries and art and craft shops. And of course, that brewery, (tel: 01 453 6700 ext 5155) founded over 200 years ago by one Arthur Guinness is open to visitors, where the black liquid can be savoured. Just off Smithfield, which has a wonderful Sunday Fair, is Bow Street where you can study the history of Irish whiskey and try a drop or two (tel: 01 872 5566). The official residence of the President of Ireland – *Áras an Uachtaráin*, is located in The Phoenix Park. Also in the Park, are playing fields, a polo ground, zoological gardens, the Ordnance Survey Office, United States Embassy, and a magnificent herd of deer, all within a mile and a half of the city centre.

For boat owners who need those bits and pieces, B.J. Marine is on the South Quay visible from the pontoon and nearby is Windmill Leisure and Marine in Windmill Lane.

Restaurants, pubs and night spots:

Dublin has a very wide range of restaurants, representing many ethnic strains. This is just a selection:

Jury's Inn, just across the road from the pontoons, has a bar serving food and a reasonably priced restaurant, tel: 01 607 5000.

A short walk away is the Financial Services Centre – Ireland's answer to Wall Street (and Threadneedle Street). There is a grocery and launderette in the complex. **The Harbourmaster Bar** lodged in what used to be the Dock Offices, serves bar food all day and there are two restaurants; upstairs is slightly posher, serving a wide range of dishes, tel: 01 670 1688.

Bewley's Oriental Cafes, Grafton Street and Westmoreland Street. A famous Dublin institution, open all day, from breakfast onwards, serving full meals as well as snacks, pastries, soups – renowned for its excellent coffee.

Elephant and Castle, Temple Bar, a bustling, lively restaurant serving pizzas, tortillas, chicken, steaks (they do not take bookings).

Kilkenny Shop Restaurant, 6 Nassau Street, set in an excellent shop for Irish crafts, wide menu, tel: 01 677 7066.

Imperial Chinese, 12a Wicklow Street, off Grafton Street, very good Cantonese, tel: 01 677 2580.

The Bad Ass Cafe, Crown Alley, Temple Bar, an airy cosmopolitan stop for pizzas and steaks, tel: 01 671 2596.

La Stampa, 35 Dawson Street, upmarket Italian, prices on the higher side, tel: 01 677 8611.

Pubs:

Well, where to start!

Davy Byrnes, Duke Street off Grafton Street, classic nineteen thirties decor, updated in places, good pub grub. Immortalised in the novel *Ulysses*.

Doheny and Nesbitts, 5 Lower Baggot Street, populated by civil servants, politicians, journalists, all intent on solving the country's problems.

McDaids, Harry Street off Grafton street, lively, good talk.

Mulligans, 8 Poolbeg Street, off Tara Street, a very old pub, good chat.

The Porter House, 16 Parliament Street, Temple Bar, a must for ale buffs. Superb beer, lager and stout brewed on premises, an incredible range of bottled beers from many countries, food served all day.

Maguires, 1 Burgh Quay, O'Connell Bridge, another pub with a micro brewery on the premises, serves food.

If you would like to keep drinking 'til the wee hours try:

Break for the Border, Lower Stephen Street (a block from Dame Street) or **Rí Rá** in Dame Court.

Refer to map on page 34.

Dun Laoghaire

Latitude: 53°18'.01N; Longitude: 06°08'.50W;
Charts: Admiralty 1415 1468; Imray C61 C62

Care must be taken in this area as high speed catamaran car ferries travel in and out of Dublin and Dun Laoghaire, reaching speeds of 45 knots. The entrance to Dun Laoghaire is clearly marked by lighthouses at the ends of the embracing east and west piers. A 600-berth marina (tel: 01 667 5585) is planned to be in operation in Spring 2001, with the entrance to the north west, and care should be taken until construction has been completed. Port and starboard hand buoys mark the marina fairway entrance and there are leading lights on the breakwater ends. The pontoons will have water and power and an amenity block will house toilets and showers. Until the marina is open, or as an alternative, visitors can call any of the yacht clubs listed below on channel 37 (0930 to 2000 hrs), identify themselves, and request a temporary mooring. These are generally available as some Dun Laoghaire mooring holders will themselves be away cruising during the summer months. All yacht clubs have pontoons (Motor Yacht Club is prone to silting), where visitors can make fast, take on water, refuel or do repairs. It is usually possible with notice to obtain a lift. The Royal St. George Yacht Club has the largest lifting capacity in the harbour – 12 tons.

The magnificent man-made harbour, one of the largest in Europe, was constructed between 1817 and 1821. The granite stone used, was hewn out of nearby Dalkey Hill and the great scar left by the quarrying is still clearly visible. The stone was brought down to the site by a railway, the path of which can still be followed, locally known as 'the metals'. The original name Dun Laoghaire, in Irish – The Fort of Leary, was restored following Ireland's independence in 1921. From 1821 it had been known as Kingstown to commemorate the visit of George IV in that year. Also marking that event is an eccentric monument, perched on four granite balls, subject of many ribald remarks, and still visible on the seafront.

With the construction of the harbour came the railway from Dublin, one of the world's first. The harbour became the terminal for the Royal Mail Steamer Packets plying to Holyhead in Wales. Locals still refer nostalgically to the modern high speed ferries as 'the mail boat'. Construction of the Royal Irish and Royal St. George Yacht Clubs followed, in classical style, with the National some time later. The number of clubs was raised to four with the construction in the 1960s of the Motor Yacht Club on the West Pier.

Dun Laoghaire has become a major yachting centre with the annual regattas now notable social events. In 1908, Sir Thomas Lipton a member of The Royal Irish Yacht Club, issued a challenge to the New York Yacht Club to compete for the America's Cup in his yacht *Shamrock IV*. The challenge foundered as Sir Thomas could not agree to the condition that he was required to sail his yacht across the Atlantic – 'on her own bottom'. However, he was a great sportsman and completed his fourth and last America's Cup challenge with *Shamrock V* in 1930. And it was from the Royal Irish Yacht Club in 1923 that Conor O'Brien left to sail around the world in his ketch *Saoirse*, built in Baltimore, Co. Cork. A large crowd turned out to greet him on his safe return in 1925. It was the first global circumnavigation by an Irish yacht.

The harbour is home to the Dublin Bay 24, a racing sloop, designed and built in Scotland in 1938 but not launched until 1946 after World War II and still racing. Its sister, the Dublin Bay 21, none of which remain alas, was originally a top sail gaff-rigged cutter, but converted to Bermudan rig. Both these fine yachts were designed by Alfred Mylne from the Isle of Bute. Another Dun Laoghaire resident is the Water Wag. This 14-foot clinker planked, gaff-rigged dinghy, dates from the 1870s and claims to be the oldest one design class in the world. They have been beautifully maintained and still engage in regular cut-throat racing. Then there are the 23-foot Glens, built by Clapham's Boatyard in Bangor and still competing after 50 years. Indeed on a Saturday afternoon and on Tuesday and Thursday evenings, up to 200 keel boats and dinghies can be seen racing in the bay. The major organiser is Dublin Bay Sailing Club, whose racing buoys are dotted around the bay during the season. They are labelled and one originally called *Pilot*, had to be renamed at the request of Dublin Port. It appears ships used to anchor alongside it awaiting the pilot boat from Dublin to guide them in! It was renamed *Poldy*, after a character in *Ulysses*. Other harbour users include the seascouts, sail training schools and St. Michael's Rowing Club. This latter body organises races in skiffs. They date from the days when pilots rowed out to incoming ships in these boats – whoever got there first won the piloting contract. Many ports in Ireland have skiff racing fleets.

Dun Laoghaire is the administrative seat of the Dun Laoghaire Rathdown County Council, whose headquarters are in the old Town Hall, an imposing Victorian building, with a landmark clock and bell tower. The two other spires visible are those of the Mariners Church, and the Catholic Church which was destroyed by fire in the 1960s. A new church was built but the old spire survived. The Mariners Church houses the Maritime Institute of Ireland and the National Maritime Museum. Here the longboat from the French frigate *La Resolue*, one of the fleet which entered Bantry Bay in 1796 is on display. She was painted in the red, white and blue of the French Tricolour, and although faded, the colours are still clearly visible. There are many other artefacts describing Ireland's maritime role both in wartime and peace.

Near the Mariners Church is the magnificent three sided-statue of Christ the King, by the Irish American sculptor Andrew O'Connor. A group purchased it and presented it to the people of Dun Laoghaire for erection on a suitable site. Considered avant garde at the time, the ascetic Catholic Archbishop of Dublin,

John Charles McQuaid, would have none of it. His view prevailed, and it was unceremoniously hidden out of sight in the garden of a private residence and subsequently in the grounds of his palace. It was a not an uncommon sight for O'Connor admirers to borrow a ladder to leg it over the wall and get a glimpse. After His Grace was called to his reward and the coast was clear, the statue was unveiled in 1978 on this fine site overlooking the harbour.

Along the east coast of Ireland, yachtsmen will have noticed the many squat cylindrical towers on headlands. These are the Martello Towers, erected in the early nineteenth century, by the British authorities, as fortifications against a possible Napoleonic invasion. Some have been converted to living quarters and are still in use. The most famous of these is at Sandycove a mile or so south of Dun Laoghaire, beside the Forty Foot bathing place. Here James Joyce lived for some time, while he taught in a nearby school. Another resident was Oliver St. John Gogarty, medical doctor, sportsman and writer. The tower now houses the James Joyce Museum, which is open daily. It was from this tower that Joyce on the sixteenth of June 1904, looked out over the 'snot green sea' and started his journey to and through Dublin via Sandymount Strand. His wanderings inspired his novel *Ulysses*, and that day is celebrated every year as 'Bloomsday', named after the book's central character – Leopold Bloom. There are now annual 'Bloomsday' celebrations. A short walk from the tower is Fitzgerald's Pub, which is dedicated to Joyce, and has some wonderful stained glass commemorating *Ulysses*.

Another literary connection can be located on the East Pier, where the anemometer is familiar to walkers and yachtsmen alike. Samuel Beckett, born and raised near Dun Laoghaire, was very familiar with the area, and often alluded to it in his works. He wrote in *Krapp's Last Tape*: 'great granite rocks, the foam flying up in the light of the lighthouse, and the wind-gauge spinning like a propeller, dear to me at last...'. This is engraved on a plaque on the lower pier, below the anemometer.

The 'Forty Foot' survived for many a year as a male only bastion of bathing. In the late nineteenth century a beady eyed policeman, espied the local gentry, bathing 'in the nip', long before modern naturist movements surfaced. Brought before the beak, they pleaded their case that as this was a resort visited only by men, therefore they were not shocking anyone. The magistrate was convinced and a compromise was agreed, and the result is still displayed by notices which state: 'Bathing costumes must be worn after 9am'! Females are since reported to have breached this male preserve, but that is another story.

Dun Laoghaire is a major residential and shopping suburb. It has been a popular holiday destination and has a fine promenade. The harbour piers attract crowds of strollers and serious walkers, all year round. It has a wide range of shops, pubs, restaurants, banks and a Post Office all within walking distance of the harbour. There is a DART station and the frequent electric trains travel south to Bray and Greystones in Co. Wicklow and north through Dublin and on to Howth and Malahide. There are also frequent bus services to Dublin and neighbouring towns.

Restaurants

The Yacht Clubs welcome visiting yachtsmen to use their facilities. Booking is required for meals and a jacket and tie for gentlemen and dresses or trouser suits for ladies are required in club dining rooms. (The Dun Laoghaire Motor Yacht Club, tel: 01 280 1371; National, tel: 01 280 5725; Royal Irish tel: 01 280 1559; Royal St. tel: George 01 280 1811).

Restaurant na Mara, located in the imposing old entrance to the railway station. Renowned for its fish, pricey, tel: 01 280 6767.

Fire Island, Lower Georges Street, colourful decor, dine here in Latin American style, your choice of steak grilled the Argentinian way, tel: 01 280 5318.

The Red Onion Cafe, 60 Upper Georges Street, inventive menu eg goat's cheese in pastry, tel: 01 230 0275.

Jewel in the Crown, 54 Upper Georges Street, Indian Tandoori, tel: 01 284 5480.

de Selby's, 17 Patrick Street, wide selection menu, tel: 01 284 1761.

Yung's Chinese, 66b Upper Georges Street, slightly expensive, tel: 01 284 2156.

Shakira Tandoori, Lower Georges Street, good Indian, tel: 01 280 0923.

The Royal Marine Hotel houses the **Powerscourt Restaurant** or for bar food try **Toddy's**, tel: 01 280 1911.

The Punjab Balti, 27a Lower Georges Street, takeaway only and much praised, take some of their delicacies back to the boat for a pleasant evening afloat, tel: 01 284 4984, 01 280 4376. They also deliver.

Pubs

Walters, 68 Upper Georges Street, a tastefully decorated pub with restful decor, restaurant attached.

Scott's, Upper Georges Street, for pub grub.

Farrel's in the Shopping Centre, Marine Road, good selection of pub grub with a fine view of the harbour.

The King Laoghaire Bar, 73 Upper Georges Street.

There are several cafes and bistros to be sampled including, **Cafe Ole** and **Baristra** in Upper Georges Street.

Refer to map on page 34.

Wicklow

Latitude: 52°58'.98N; Longitude: 06°02'.70W;
Charts: Admiralty 633 1468; Imray C61

Leaving Dun Laoghaire and turning south yachts can safely make the passage between Dalkey Island and the mainland before the beautiful vista of Killiney Bay opens up. It has been compared to the Bay of Naples with the aptly named Sugarloaf Mountain playing the part of Mount Vesuvius. Many roads in the area reflect the connection eg Sorrento Terrace and The Vico Road. Here some of Dublin's finest homes with many famous residents from show business and sport can be found including Bono of U2, Enya and Damon Hill. Past Bray Head and town, the lighthouse on Wicklow Head should be visible. Trains on their way to and from Rosslare can be seen threading through tunnels and travelling along the coast.

The entrance to Wicklow Harbour is clearly marked by lights on the East and West Piers. It is both a fishing and commercial harbour but fendered berths have been established on the East Pier for visiting yachts. Although giving a longer walk to the town, you will be out of the way of trawler and coaster traffic. Local yachts moor towards the West Pier.

The name Wicklow derives from its use as a haven for Norse sea-rovers. They called it Wykynlo or Viking's Loch. In Irish Wicklow is Cill Mhantáin or Mantan's Church. It is associated with the O'Byrnes who harassed the English settlers from their strongholds in the nearby mountains. Wicklow played a major part in the rising of 1798.

Today it is a busy town, and is a good starting point for those who want to explore the nearby countryside. There are beautiful mountains, lakes and forests with stately mansions including the famous Powerscourt demesne, and ancient monastic settlements including Glendalough. In Fitzwilliam Square a monument commemorates Captain Robert Charles Halpin of the *S.S. Great Eastern* which laid the first cable to connect Ireland and the United States.

Wicklow Gaol has been renovated, not to house recalcitrant sailors, but as a museum. The grim and harsh realities of early prison life are displayed, along with a description of the role it played during the famine years and the 1798 rebellion. As it was a holding centre for almost 50,000 Irish people being deported to Australia as convicts, there is a special section devoted to them and the part they played in the development of Australia. The gaol, at the southern end of the town, is well signposted and is open from 1000 to 1700 daily.

Wicklow has a busy sailing club which welcomes visitors. They have showers (a contribution to expenses is gladly acknowledged); the bar opens at 2030. Wicklow S.C. hosts the famous biennial (even years) Round Ireland Yacht Race which starts and finishes off Wicklow Head.

Restaurants and Pubs:

Try **Phil Healy's**, Fitzwilliam Square, for good bar food (1200 to 2200) and drinks. There is also a restaurant called Phil's attached.

The **Bakery Restaurant and Cafe**, Church Street, for a range of fish, fowl and steaks. They also run a coffee shop next door, tel: 0404 66770.

The **Bridge Tavern**, and **Crow's Nest Bar**, Bridge Street, serving bar meals from 12 noon to 2000 (Captain Robert Halpin mentioned above was born in the building in 1836). They host regular music gigs. There is a quayside beer garden, pleasant in fine weather, overlooking the harbour and the Leitrim river, tel: 0404 67718.

Rugantino's, South Quay, which has a riverside balcony, offers Mediterranean/Irish cuisine, good selection of meat and fish. Lunch Wednesday to Sunday, dinner Tuesday to Sunday, tel: 0404 61900.

The Old Forge, Abbey Street serves bar food. The **Blacksmith Restaurant** is upstairs, serving lunches and dinner, tel: 0404 66778.

Hannahs in the Main Street offers good breakfasts and cafe food.

Refer to map on page 35.

Arklow

Latitude: 52°47'N; Longitude: 06°08'.20W;
Charts: Admiralty 633 1468; Imray C61

Leaving Wicklow and turning south, the very popular beaches of Brittas Bay come into view. The masses of trailer homes confirm that this is a major holiday home destination for Dubliners.

Passage should be made to the west of the Arklow Bank, the first of a series of banks between here and Greenore point (Rosslare). Arklow has lights at its harbour entrance on the ends of the north and south walls. Entrance is very difficult without an engine, and a swell develops in easterly winds. Watch out for commercial and fishing traffic. Having rounded into the river the harbour is to the left. The marina (40/50 berths) is located opposite the harbour – keep straight on, holding to the right, where the marina entrance will be visible; telephone contact is Arklow Shipping, 0402 39901 or mobile 087 237 5189.

Marina charges are £11 per day up to ten metres and rising a pound per metre up to £16 for a 15 metre boat. Weekly rates are £66 for a boat up to ten metres and rising £6 per metre up to £96 for a 16 metre boat. There is water and power on the pontoons (£2 power charge per day). Showers and toilets are close by. Fuel not yet available (there is a filling station about a half mile away).

An alternative to the marina is a berth in the inner harbour basin about 300 metres upriver on the port side. Tie up along-side other boats at the east wall. Don't cross the basin diagonally at low water. There is enough depth if you follow the line of the north and east walls. The latter culminates in the Lifeboat House. The berthing master is the engineer on the lifeboat and their showers are made available to visiting sailors. There are no other facilities. The charge for a ten metre yacht is £7.00 per day and every third day is free.

Arklow is another town name derived from the Vikings who welcomed the shelter it gave. In Irish, the town is known as Inbhear Mór or Large Estuary. From being a Norse settlement during which time the port was developed, the town became an Anglo-Norman possession. It was ruled by the Butler branch of the Ormonde family who constructed a fortress. This was demolished in the Cromwellian invasion in 1640s. Subsequently the Irish insurgents in the 1798 rebellion against English rule suffered a crushing and crucial defeat. One of the insurgent leaders, the legendary Father Murphy was killed in the battle, and a monument in his memory, can be seen in Upper Main Street.

During the First World War a major munitions factory, Kynoch's was located behind the North Quay. It was almost totally destroyed in a disastrous explosion, with loss of life, and closed in 1920, with many workers emigrating to South Africa to work for a sister plant of Kynoch's. In recent times Arklow developed industrially around a fertiliser plant, incredulously located in the beautiful nearby Vale of Avoca. In the 1930s a pottery was established, and the Stoke-on-Trent type bottle kilns were a familiar landmark. Alas the pottery is now closed but its buildings are still standing on the inner harbour quays. Another famous industry, now also closed, was Tyrrel's Boatyard. Here Francis Chichester's *Gypsy Moth III* was built, as was the Irish Sail Training Brigantine *Asgard II*, a well known vessel in Irish waters and in many parts of the world.

Arklow is a commercial port, taking small coasters. It is the headquarters of Arklow Shipping Ltd., which operates one of Ireland's largest commercial fleets. Their modern office building is on the North Quay, and beside is the clubhouse of Arklow Sailing Club. There is a small fishing fleet. There still is an active boat repair business and the inner harbour has a hydraulic lift which can handle trawlers and similar vessels for overhaul.

Arklow has a rail connection south to Rosslare and north to Dublin. There are also bus connections. To reach the town, walk up the North Quay to the clearly visible 'nineteen arches' bridge carrying the road to Dublin. Turning left the junction with Main street is clearly visible. The town has the usual facilities within easy reach. The Post Office, supermarkets and banks are all located in the Main Street. The Tourist Office is off Upper Main Street. There is an interesting maritime museum located in St. Mary's Road.

Arklow has a selection of restaurants, pubs and takeaways; most pubs serve barfood and many have a restaurant attached; they are all located in Main Street unless otherwise stated:

The **Bridge Hotel**, (between the bridge and Main Street) has a bar with much nautical memorabilia. Meals are served, tel: 0402 31666.

Rippley's, for bar food, carvery and evening meals, tel: 0402 32683.

Christy's, **Conservatory Restaurant** attached, tel: 0402 32145.

Kitty's, **Loft Restaurant** attached.

Birthistles with **Lazy Lobster Restaurant**, for seafood, tel: 0402 32253.

The **Riverwalk Restaurant** is aptly named as it is on the River Walk, off the Main Street and overlooking the Avoca River, tel: 0402 31657.

The **Nineteen Arches** in Lower Main street has music sessions at the weekend as has **The Mary B Lounge**.

For takeaways try the Chinese in Castlepark off Upper Main Street or the **Roma** for pizzas in Upper Main Street.

Refer to map on page 35.

Wexford

Latitude: 52° 20'N; Longitude: 06° 27'W;
Charts: Admiralty 1772 1787; Imray C 61

Leaving Arklow and turning south are the Glassgorman Banks which should be avoided. Passage should be made through the Rusk Channel. The entrance to Wexford Harbour is marked by a buoy about eight cables ESE of Raven Point. Wexford no longer has commercial traffic, but it is used by trawlers. During the summer months the channel, which changes constantly, is marked by orange buoys. A 60-berth marina has been opened at the Crescent on the pier waterfront. It is advisable to ring Wexford Harbour Boat Club on 053 22039 before attempting an entrance to the harbour.

Wexford derives from the name 'estuary of the mudflats', which the Norse called it, establishing a flourishing trading post in the ninth century. It eventually fell into Anglo-Norman hands, who established a walled city – today the narrow winding streets reflect those origins. Sacked by Cromwell in 1649, it became the centre for the uprising of 1798.

In Westgate, The Westgate Heritage Centre offers an audio-visual presentation on the origins and history of Wexford. Also on display is a section of the wall of the old fortified town.

Wexford is renowned for its opera festival which runs from mid-October to early November. Sailors will undoubtedly be in dry dock by then, but opera lovers among them, might like to return to this unique event, which stages the lesser known or forgotten works, often of the most famous composers. There are many supporting musical events. It has a relaxed atmosphere with black-tied men and gowned ladies spilling out on to the street during opera intervals at the Theatre Royal, to enjoy a drink. Impromptu sessions can often be found in the local hostelries.

For eating the two main hotels offer a range of meals:

Talbot Hotel, **Paul Quay**, **The Slaney Restaurant** and **Trinity Bar** also serving food, with jazz sessions on Thursdays, tel: 053 22566.

Whites Hotel, North Main Street, has **Harper's Restaurant and Bar**, tel: 053 22311.

For a range of Italian dishes in pleasant subdued surroundings, try **La Dolce Vita**, Westgate, tel: 053 23935.

La Riva, The Crescent, at the top of the stairs, seafood, lamb and pasta await, tel: 053 24330.

Mirabeau, Anne Street, another intimate spot for fish and game, tel: 053 21777.

Mange 2, 100 South Main Street, simple decor, serving an interesting selection, tel: 053 44033.

Robertino's, on North Main Street for pastas.

Into the Blue, North Main Street, open all day, tel: 053 22011.

Wexford also boasts a fine range of pubs:

The Tack Room, North Main Street, excellent pub grub served until 2030.

Archer's, Redmond Square, renowned for its lunch carvery.

Centenary Stores, Charlotte Street, a pub with a mellowed interior untouched for many years.

Westgate Tavern, Westgate, another pleasant pub with a relaxed atmosphere serving food.

On Main Street South, look for **Tim's Tavern** and **Simon's Place** both serving pub grub.

If your legs can only carry you to the seafront, there are a few bars including **O'Beara's – The Wren's Nest**, Custom's House Quay, which has traditional music sessions.

Refer to map on page 36.

Locations

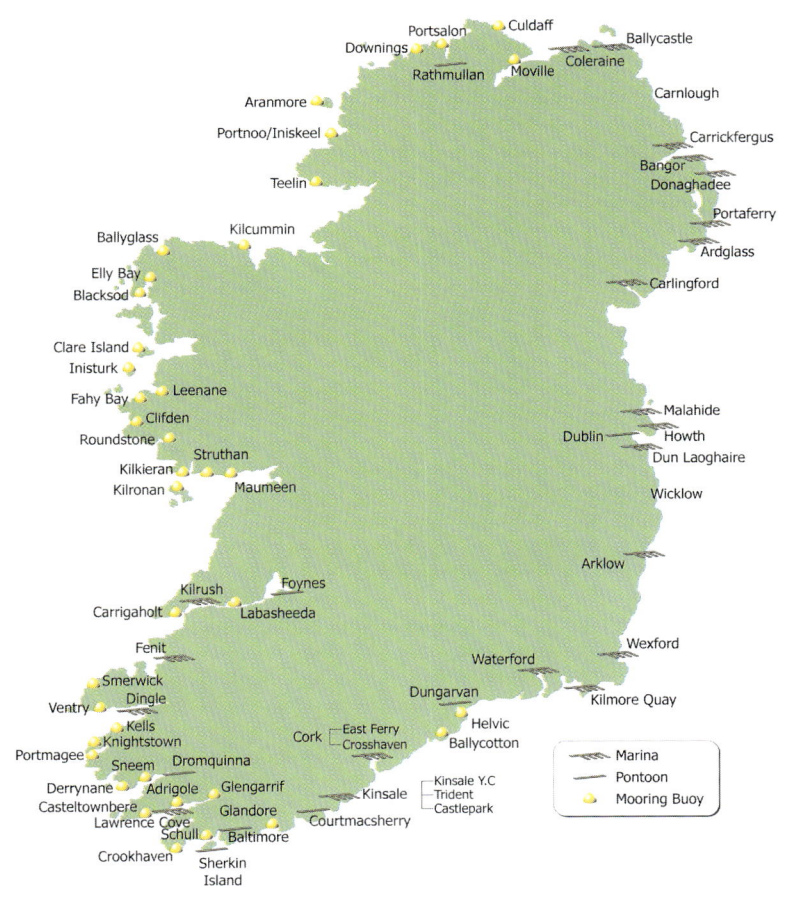

Killiney Bay

Maps

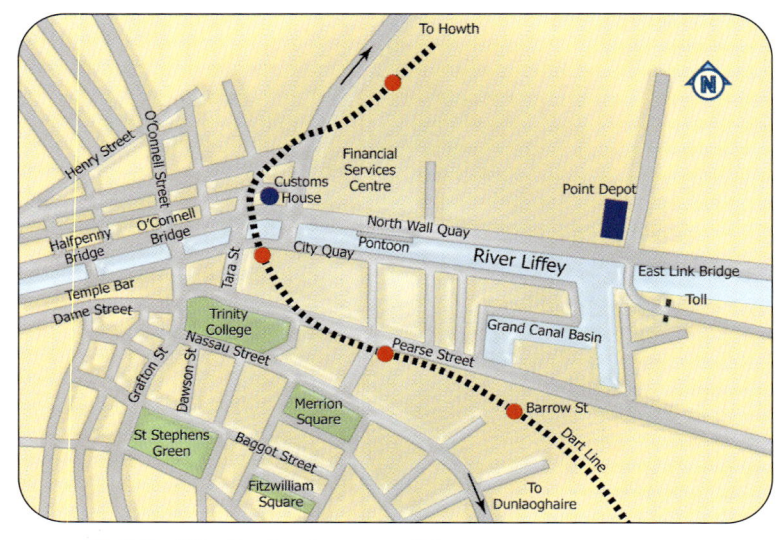

Dublin City Centre (see page 18)

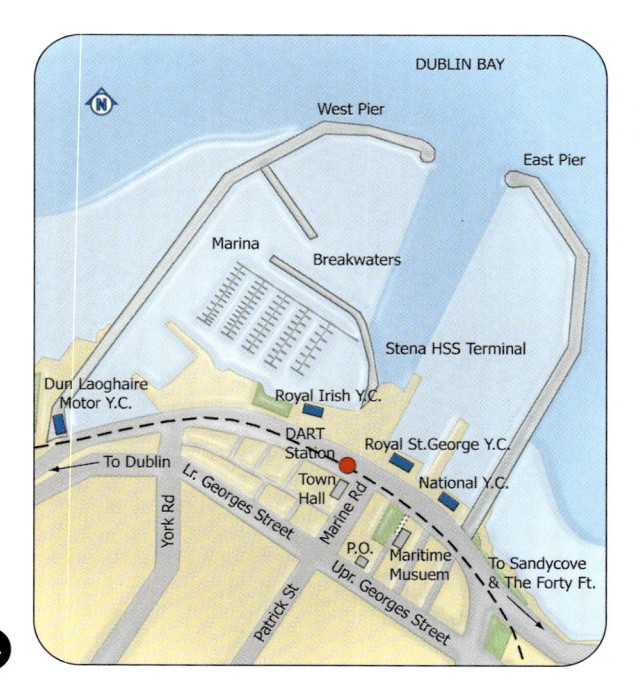

Dun Laoghaire (see page 23)

Maps

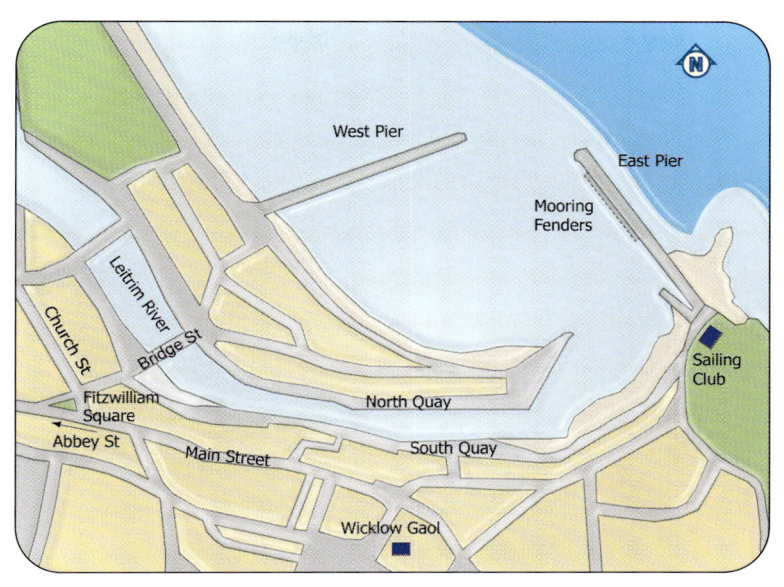

Wicklow (see page 28)

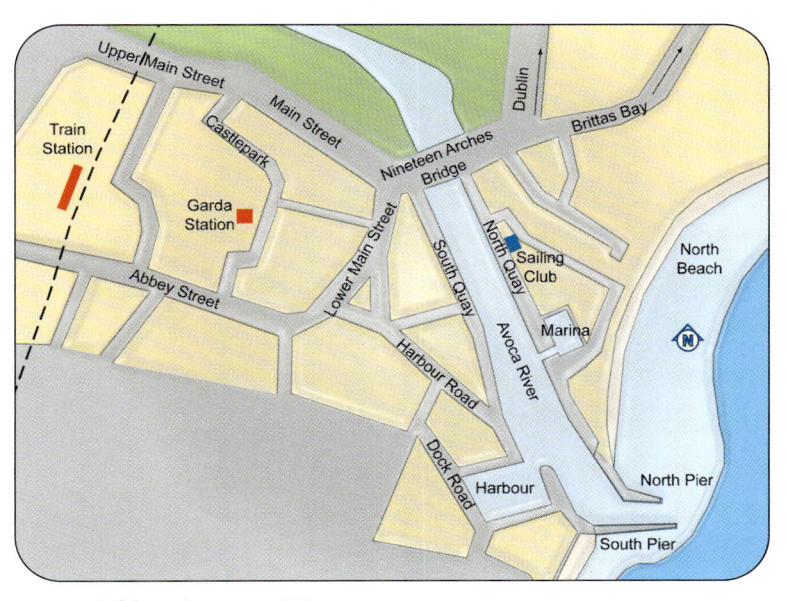

Arklow (see page 29)

35

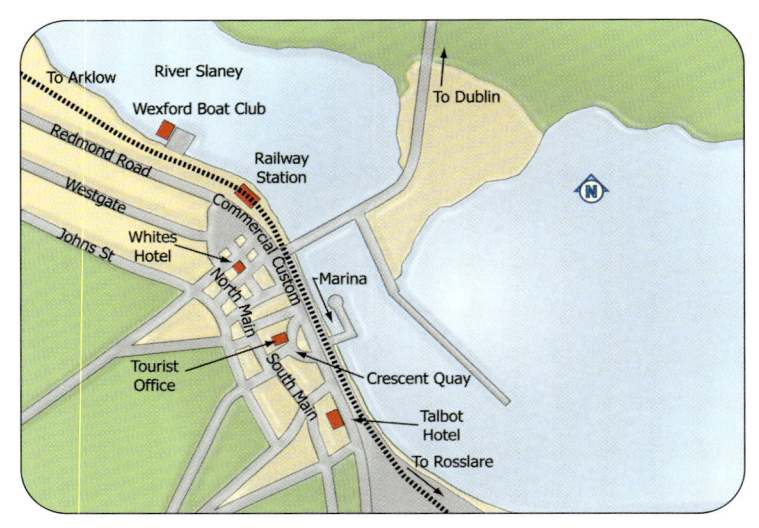

Wexford (see page 31)

Kilmore Quay

Latitude 52°10'N; Longitude 06°35'.50W;
Charts: Admiralty 2049 2740; Imray C61 C57

Leaving Wexford and turning south, a series of banks is clearly marked. At the end of these banks lies Rosslare, a busy port for conventional and high-speed ferries heading for Pemroke and Fishguard in Wales, and Cherbourg in France. The Tuskar Rock lighthouse lies about six miles off the south east 'corner' of Ireland, Carnsore Point. A word of caution; strong onshore winds can produce nasty seas here, so beware. Having rounded the point watch out for The Barrells, a group of dangerous rocks. The Saltee Islands, now uninhabited come into view and form some of Ireland's most important bird sanctuaries, with shearwaters, razorbills and guillemots nesting there; seals can be seen taking the sun. Kilmore Quay is a fishing port and lifeboat station and today is a popular stopping point in an area which was traditionally short on places to tie up.

Entrance to Kilmore Quay from the west is straightforward, but consult chart or almanac to get you safely clear of the Forlorn Rock. From the east if the wind is light you can cross St. Patrick's Bridge, a gap in a submerged spit coming out from the shore. The opening is buoyed but these are small and hard to see in rough weather. Once through, do not turn right until the harbour entrance leading lights are lined up. Alternatively, approach through the Great and Little Saltee Islands, avoiding the Goose and Murroch's Rocks. Your onward trip can be planned with help of locals who will be only too willing to advise on winds and tides, and how to negotiate Carnsore Point if travelling up the east coast.

The 55-berth marina (tel: 053 29955) has power, water and holding tank draining facilities. Charges are £10 per day up to ten metres; add £1 per extra metre. For longer stays the seventh day is free. The ever helpful berthing master, Mr. Sinnot will see you safely tucked in. Toilets are on the West Pier. Contact the Fishermen's Co-op regarding fuel. Kilmore Quay is a delightful village, renowned for its thatched cottages. There is a shop, public telephone and Post Office. Showers are available in the Stella Maris Centre which also has a coffee shop and Tourist Information Office (open 0900 to 2000). There is a well stocked chandlery. A sign beside the harbour displays details of the Wexford Coastal Walks.

The village provides opportunities to eat within a short walk. All serve locally caught fish – **The Silver Fox**, a restaurant renowned for its excellent and extensive fish menu, probably the best in Ireland. Booking is advisable, tel: 053 29888.

James Kehoe's pub serves very good seafood (and other dishes) as does the **Wooden Door**. There is a fish and chipper, which is excellent, just outside the marina entrance. The proprietor is a well-known diver.

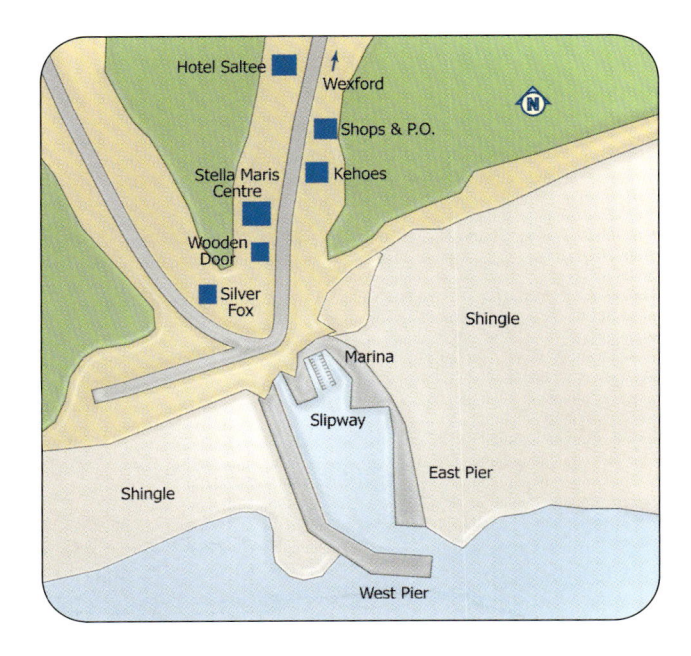

Waterford

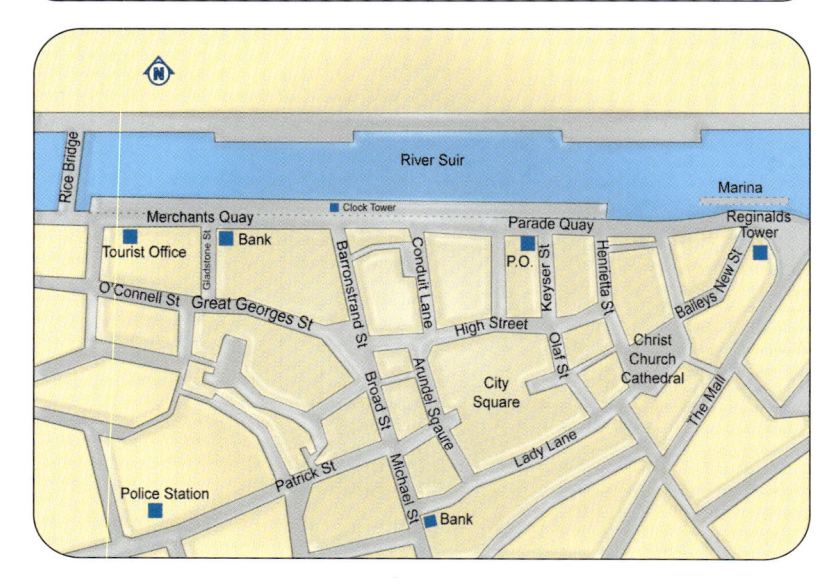

Latitude: 52°15'.50N; Longitude: 07°06'W;
Charts: Admiralty 2046; Imray C57

Leaving Kilmore Quay and travelling south west, the expanse of Ballyteigue Bay opens up. To the west is Hook Head and lighthouse (reputedly one of the oldest in Europe) marking the easterly entrance to Waterford Harbour; to the west is the fishing harbour of Dunmore East. Waterford city is situated on the River Suir, a picturesque journey of about 12 miles from the estuary entrance. Do not attempt the journey aginst the tide unless you have a strong engine. The channel takes you north past Creadan Head to the confluence of the Suir and Barrow rivers at Cheek Point. From there the channel turns south west passing the container terminal and power station. The channel then turns west passing the northerly side of Little Island. The marina lies on the south side of the river.

The marina (tel: 051 873501) has 100 berths, and can be contacted on channel 37. Daily charges are up to 18 feet – £6, 18 to 25 feet – £8, 25 to 35 feet – £10 and 35 to 45 feet – £12. There are discounts for longer stays. There is fresh water and 240 volt power on the pontoons. Fuel can be obtained at a nearby petrol station. All the usual amenities of a city; phones, banks and Post Office are all within easy reach. Showers are available at Viking House. When the marina is busy boats raft up two or three deep. This can present problems leaving due to the confined space especially when the tide is in full spate as it runs at three knots or more. In those circumstances it is wise to plan your departure for one hour either side of slack water.

Waterford is Ireland's fourth largest city, and the stalwart citizens claim that it is the oldest, with the civic motto 'Urbs Intacta'. The business sector is located mainly to the south of the river. Waterford was first settled by the Norse in the tenth century, who saw its potential, and developed the port which became a busy trading location. The Norse adopted Christianity and were nominally dependent on the King of Munster. In the twelfth century, Diarmuid MacMurrough, King of Leinster in effect annexed Waterford. But in 1170 it fell to the Norman Strongbow (who is buried in Christchurch Cathedral in Dublin). To cement his claim to the kingship of Leinster, Strongbow married Aoife, the daughter of MacMurrough in Waterford Cathedral.

Waterford remained a Catholic city loyal to the English crown. It resisted various attacks by anti-loyalists, but was eventually conquered by Cromwell. Its trade and influence declined. But the city produced many famous figures including the Franciscan scholar, Luke Wadding, Rev. Francis Hearn Professor of Rhetoric at Louvain University in what is now Flemish speaking Belgium. Another illustrious son is Thomas Francis Meagher, a Young Ireland leader during the insurrection of 1848 against English rule. Condemned to death, Meagher escaped to America, fought at Fort Sumter and in time became Governor of Montana.

Waterford was renowned for its glass making in the eighteenth and nineteenth centuries, but as with many other industries, punitive tariffs led to decline. After many years, the industry was revived in the 1940s. It became one of Ireland's outstandingly successful industries, eventually acquiring Wedgwood in England and Rosenthal in Germany. It now ranks as one of the world's foremost suppliers of high quality tableware. The Waterford glass factory is located about a mile outside Waterford on the Cork Road. Visitors are welcome to visit and watch as the glassware is blown, and a glass or decanter takes shape, and is then skillfully cut. Waterford is also home to a range of other industries including food processing and engineering. There is a major container port at Bellhaven on the north side of the river; good rail and bus connections.

Many archaeological remains, witnesses to the historical past are to be seen. Across the road from the marina entrance, at the corner of Parade Quay and The Mall stands Reginalds Tower, in turn mint, prison and military store, which now houses an exhibition. This formed part of the original city walls, the remains of which are still visible in many parts of the city. The exhibition is open from Easter to October, Monday to Friday 1000 to 1700 and Saturday/Sunday – 1000 to 1800. Admission is £1.50 for adults, children/students 60p, senior citizens and groups £1.00 and families – £4.00.

There are many notable buildings including the Church of Ireland and the Catholic cathedrals both dating from the late 1700s; the Court House, City Hall and the Georgian residence now occupied by the Chamber of Commerce in

Georges Street. The Granary, Merchants Quay, has Waterford Treasures, featuring many artefacts from the city's past. There is a charge for admission.

Waterford has an extensive range of restaurants, pubs and takeaways. The following is merely a selection:

Bewley's Cafe, Broad Street Shopping Centre – famous for their coffee.

T and H Doolan's pub in Georges Street, a timbered building, one of Ireland's oldest taverns with a good selection of bar food, folk music sessions in the evening.

The Olde Stande Pub, Michael Street has a restaurant upstairs with old world ambience, serves steaks and seafood, tel: 051 879488.

The Wine Vault, High Street, Seafood Bistro, Wine Bar and Wine Shop, tel: 051 853 444/853 777.

La Palma, Parnell Street (near The Mall), Italian specialities, tel: 051 879 823.

McDonald's, Barronstrand Street, this fast food restaurant needs no introduction.

Jade Palace, Chinese, beside Reginald's Tower, tel: 051 855 611.

Downes Pub in Thomas Hill has the distinction of serving its own whiskey – Downes No. 9.

Reddy's Bar, High Street serves pub grub.

Jordan's American Bar on The Quay, a past favourite for dockers, has retained its old charm.

McCloskey's Bistro, High Street, well praised with creative menu including speciality vegetarian dishes.

There are several hotels within easy reach; **The Tower** (beside the marina), **Adelphi Riverside Bar** with live music at weekends and **Imperial Restaurant**, tel: 051 875 801, **The Marina** (behind the Tower Hotel), tel: 051 877 222, **The Bridge** with **Crokers Restaurant and Bar**, and **Dooley's** with **The Dry Dock Bar** and **New Ship Restaurant**, tel: 051 873 531, both on Merchant's Quay.

Helvic

Latitude: 52°03'.30N; Longitude: 07°33'.10W;
Charts: Admiralty 2017 2049; Imray C57

Leaving Waterford and turning west, a panoramic coastline opens up. Inland are the verdant slopes of the Comeragh Mountains. A few miles from Waterford lies the coastal resort of Tramore. Helvic Head is the promontory at the southern entrance of Dungarvan Bay. Under the head lies the fishing harbour of Helvic. There are eight visitor mooring buoys located about 300 metres to the south west outside the harbour entrance. Once ashore a visit should be made to **Tig Muirite** or the pub of landlord Muirite. The name is in Irish as this part of County Waterford is a Gaeltacht or Irish speaking area. Nearby in the village of Ring a famous Irish College is located, where many young Irish people go to learn their native tongue.

Dungarvan

Latitude: 52°05'.15N; Longitude: 07°36'.70W;
Charts: Admiralty 2017 2049; Imray C57

From Helvic, the town of Dungarvan lies to the northern end of the bay. The entrance is clearly buoyed. However, yachts having passed through the last red and green markers, should keep to the left hand side close to the quay wall, to avoid grounding, in the direction of the Sailing Club's pontoon on Davitt's Quay, on the western side of the harbour. Visitors can berth here temporarily before checking with the Sailing Club which is about 50 metres away. They may then be given a visitor mooring. The Sailing Club has showers and a bar (check with them for times of opening). There is a £10 charge for mooring (or pontoon if available) per three days; visitors will be given a smart card for access to the club.

Dungarvan has Anglo-Norman origins, and the remains of a castle of that era can be seen. Today it is a busy town on the Colligan River.

Pubs and restaurants:

On Davitts Quay is **King John's Patisserie**, an all day cafe/restaurant serving meals and snacks until 2100. It also has takeaway food.

Next door is **Jade Palace** a Chinese restaurant, tel: 058 45370.

The Moorings and the **Anchor Bar** are conveniently located on the Quay, and serve good food.

The Mill Restaurant is in an old mill house near the Sailing Club, serving steaks, lamb and fish, tel: 058 45488.

The Tannery, Quay Street, which is a Bib Gourmand, a Michelin category – good food at modest prices, tel: 058 45420.

Merrys Bar and Restaurant, Lower Main Street, featuring seafood and game in season, tel: 058 41974.

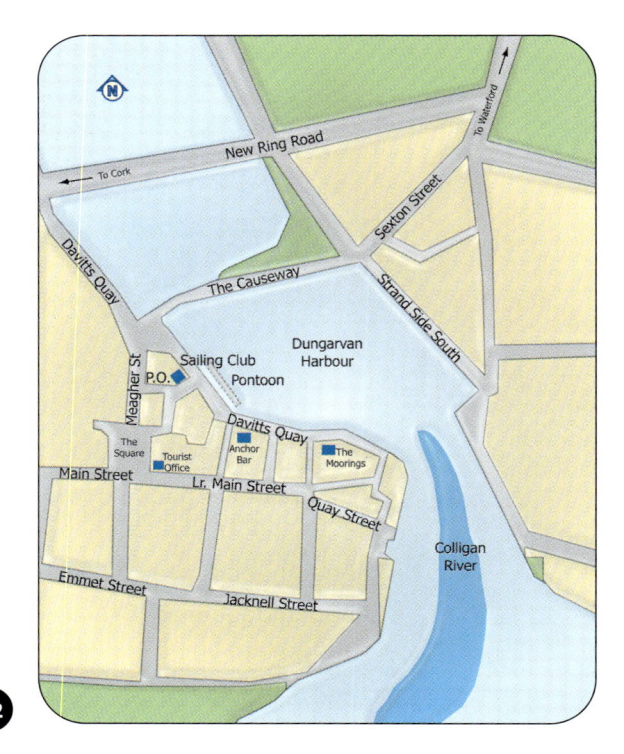

Dungarvan

Ballycotton

Latitude: 51°49'.50N; Longitude: 08°W;
Charts: Admiralty 2049; Imray C57

Leaving Dungarvan Bay and turning west first Mine Head, then Ardmore Head, next Knockadoon Head at the eastern entrance of picturesque Ballycotton Bay come into view. Ballycotton Island is recognisable by its prominent lighthouse. To the west and tucked in behind the island lies Ballycotton a pleasant East Cork fishing village. There are six visitor mooring buoys about 100 metres to north of the harbour entrance. The harbour is home to a busy fishing fleet and the local lifeboat. There is a Post Office and usual range of village shops.

The village is well served with restaurants and pubs:

The Blackbird of Ballycotton (An Ian Dubh), a pleasant hostelry with music at weekends.

Spanish Point Restaurant, tel: 021 646 177.

Bayview Hotel, tel: 021 646 748.

Pubs include **Inn by the Harbour**, **Fawcett's**, serving pub grub; **Schooner Tavern**.

About two miles distant is Shanagarry, home to the Stephen Pearce Pottery. Nearby is the famous Ballymaloe House Hotel. This has a world famous cookery school at Shanagarry and the Allen family has made Ballymaloe synonymous with excellent cuisine using locally harvested ingredients. Darina Allen is a regular contributor to Irish TV cookery programmes and author of several cookery books, which have an international reputation.

Cork Harbour

Charts: Admiralty 2049 1777 1773 1765; Imray C57

About ten miles south west of Ballycotton, the entrance to mighty Cork Harbour opens up. Cork, Ireland's third largest city, after Dublin and Belfast, is rightfully proud of the fact that it possesses one of the world's finest natural harbours. The Harbour is entered between Weaver's point to the west and Roche's Point to the east. It is busy with ferry and commercial traffic. Just past the entrance and to the west is Crosshaven, Cork's yachting centre described in detail below. To the east can be seen the oil refinery at Whitegate. Further into the harbour to the west is the ferry terminal of Ringaskiddy, with its industrial plants. Within the harbour there are three islands. The largest is Great Island which has ship repair facilities in Rushbrooke at the west side. Cobh, formerly Queenstown, is the island town. Haulbowline is the headquarters of the Irish Naval Service, where there is also a steel processing mill. Britain's last presence within the Republic was here with the Royal Navy leaving in 1938. Spike Island for long a penal settlement is still used to accommodate young offenders.

The Harbour will always be associated with the great transatlantic liners, which regularly called until the 1960s. But there are sad memories. It was from Cobh that the *Titanic* sailed on her ill fated voyage in 1912. The *Lusitania* was torpedoed and sunk off the Old Head of Kinsale, in 1915, about 25 miles south west of the harbour entrance.

The visiting sailor has a choice of two marina locations, East Ferry and Crosshaven.

For more information on Cork City, see entry on Crosshaven.

East Ferry

Latitude: 51° 51'.70N; Longitude: 08°12'.50W

There is a 100-berth marina located in the picturesque East Passage entered between Gold Point and Marlogue Point with water and power on the pontoons. The marina has its own restaurant, - **The Marlogue Inn and Bar**. Showers and toilets are located in the rear of the compound. Visitor rates are £8 per day for a 25-foot boat, £10 for 30 feet and add £2 for each additional five feet. A week's stay is discounted to five days.

The town of Cobh is about four miles by road. If the marina host, Mr. Butler, is not there to offer a lift into town, a taxi can be called from the phone in the restaurant. Cobh has an attractive waterfront, with a wide range of shops, banks, restaurants and pubs. Above the town towers the gothic cathedral of St. Colman, famous for its carillon of bells. In the restored Victorian railway station, which connects Cobh with Cork City, is 'The Queenstown Story', a centre where the maritime history of Cobh is relived. There is a monument in the town centre to the memory of those lost in the *Lusitania* disaster, many of whom were buried in local graveyards. A smaller memorial nearby commemorates the connection with the town of the ill-fated *Titanic*.

On the western end of the seafront stands an elegant building, once the clubhouse of The Royal Cork Yacht Club, reputedly the world's oldest yacht club, founded in 1720. The clubhouse closed in the 1960s, and was amalgamated with the Royal Munster Yacht Club in nearby Crosshaven, which then adopted the Royal Cork name. Being so old and Cork-based, the club does not stoop to calling its head sailor a mere commodore, but he/she quite rightly carries the title 'Admiral'. The original clubhouse fell into serious neglect and disrepair, but has recently been rescued and is being refurbished.

Cobh has a reasonable range of pubs and restaurants along the seafront.

Restaurants:

Bistro, tel: 021 481 1237; **Jim's Place** (lunch); tel: 021 481 1497; **Mimmo's**, tel: 021 481 1343; **Peninsula**, Chinese with takeaways, tel: 021 481 3791; **Trade Winds**, tel: 021 481 3754.

Pubs, many of which have a nautical theme in their name: **The Ship's Bell**, **Welcome Inn**, **The Admiral**, **Rob Roy Bar**, **The Voyager**, **Lusitania Bar**, **The Rotunda**.

Crosshaven

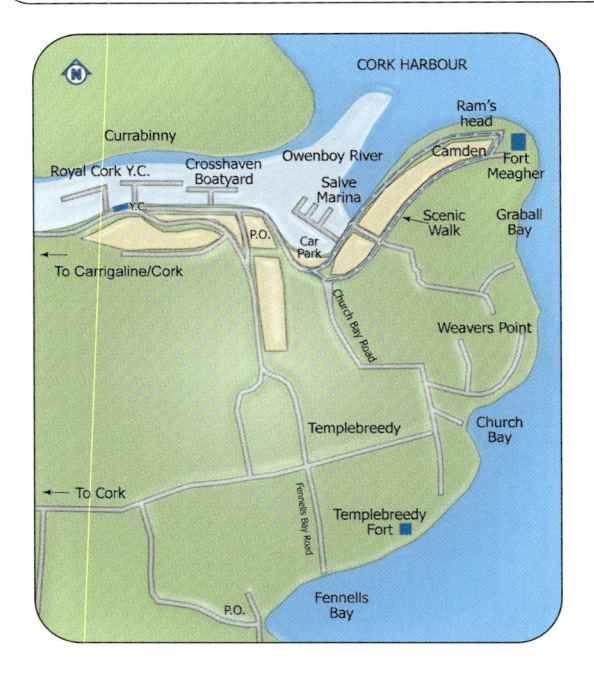

Latitude: 51°48'.30N; Longitude: 08°18'.90W (Royal Cork Y. C.)

Crosshaven is a village situated on the estuary of the picturesque Owenboy River. Across is the wooded hillside of Currybinny. Tradition has it that in 1580 Sir Francis Drake, pursued by the Spaniards took refuge just up the river. Not aware of his location, the Spaniards sailed up to Cork, and down again empty handed, leaving without having found their prey. Today the point of refuge is still called 'Drake's Pool'.

A walk of a little over a mile brings one to Ram's Head where there is a magnificent view of the harbour, and of ships passing through the channel below. Here also is Fort Meagher, one of the original fortifications of the harbour and a fine example of a coastal artillery fort. Across the water is Roche's Point with its lighthouse, marking the eastern harbour entrance, and the Met Eireann weather station from which conditions are reported for inclusion in forecasts.

There is a bus service to Cork city about 12 miles away. Taxis are also available – enquiries can be made at any of the marinas. Cork is a city whose centre has changed little over the years. Recent building and infrastructure development has largely taken place in the surrounds, and many of the streets are still faithful to their origins. The river Lee flowing through the city divides to create an island in the city centre, so it is a city of bridges and river views. Commercial shipping traffic berths right in the centre. Probably its most famous building is St. Finbarre's Cathedral. Nearby is St. Ann's church with its famous clock that rings out the time over Shandon. As a true cockney is born within sound of Bow Bells, the real Corkman was born to the sound of the 'bells of Shandon, that sound so grand on, the pleasant waters of the River Lee'. An equally famous song, much loved by the citizenry, praises 'the banks of my own lovely Lee'. Another notable building is the City Hall. Cork possesses all the amenities of a large city – theatres, cinemas, restaurants, museums, university, interesting markets. There is a fine art collection in the Crawford School of Art. Cork has an international airport, with frequent services to Dublin, the UK and continental Europe. There are bus and rail connections to all parts of Ireland. There are car ferry connections from Ringaskiddy in the harbour to Swansea in Wales and Roscoff in France (once weekly). For more details contact the Tourist Information Office, tel: 021 427 3251.

Crosshaven is a major boat and yachting centre. Boat lifting, overhaul, repair and maintenance services are available, as is brokerage. McWilliam Sailmakers are located about four miles distance, off the road to Cork.

Crosshaven is home to The Royal Cork Yacht Club, justifiably renowned for its hosting of the biennial (even years) Ford Cork Week. This now ranks as the largest yachting event in Europe with around 800 boats taking part. During this event, a tented city appears in the yacht club grounds, providing backup and entertainment facilities for the competitors and their guests. It is also associated with the 1720, a fast racing yacht, named after the club's foundation date.

The Cork yachtsman is a unique breed, not playing golf during the sailing season or events. Doing so could give the impression that he or she might not be a boat owner! Crosshaven offers a choice of three marinas, Royal Cork, Salve and Crosshaven Boat Yard.

Royal Cork Yacht Club Marina, tel: 021 483 1023.
The club welcomes visitors who should preferably make prior arrangements. Visitors should tie up at the eastern end of the marina and report to the office or if closed the bar, and should produce evidence of yacht club membership. Water and electricity are available on pontoons. Showers are also available. Meals and sandwiches are available in the restaurant and bars. However, male visitors are expected to wear a jacket and tie and female visitors should be appropriately dressed after 1900 at weekends.

There are limited supplies of diesel available, but petrol is not stored. The club office will be pleased to supply telephone, fax and e-mail facilities at reasonable cost. Daily rates for visitors are: £12 up to 25ft; £14, 25ft to 30ft; add £2 for each extra 5ft. There is a discount for weekly and monthly stays. Swinging moorings can also be rented at 50% of the above rates.

Salve Marina, tel: 021 483 1145.
Just to the east of Royal Cork is the 70-berth Salve marina. Examples of daily berth rates: are £12 up to 30ft in length; £15, 31ft to 40ft; £17, 41ft to 50ft. Weekly rates are charged at five days. Swinging moorings are available at daily rates: £8, 30ft to 40ft and £12, 41 to 55ft. Enquire about rates for lengths above these. There is water and electric power on the pontoons. Showers and hull, rigging and engine repair services are all available on shore. Fuel is available.

Crosshaven Boatyard Co. Ltd. tel: 021 4831 161.
Close by to the east again is Crosshaven Boatyard Co. Ltd. Marina with 120 berths. Visitors should call on channel 37. Example daily rates: £9 up to 25ft; £11, 26ft to 30ft and £14, 31ft to 35ft. Weekly rates are charged at five days. Showers are available, and also fuel. There is a fully equipped boatyard, boat park with 40 tonne travelling hoist.

There is another boatyard close by - Castlepoint Boatyard where yachts and motorboats in GRP and steel are built and repaired. A full range of overhaul and maintenance services is offered,

The village has a small selection of restaurants and pubs:

The Schooner Bar and Restaurant, serves excellent seafood, steaks and game, tel: 021 436 1806.

Cronin's beside the village car park offers very good pub food, and has an interesting collection of photographs tracing the maritime history of the area. In summer you can sit outside and take in the views.

Also close by on the waterfront are the **Admiral Drake**, **The Moorings**, and **Johnny's Return.**

Kinsale

Latitude: 51°42'N; Longitude: 08°31'W;
Charts: Admiralty 2053 1765; Imray C56 and C57

Leaving Cork Harbour in a southwest direction, the entrance to Kinsale Harbour lies about twelve miles away. Beyond the entrance is the Old Head of Kinsale and lighthouse. Kinsale is reached through the promontories of Chroohoge and Preghane points, leaving the Bulman Rock Buoy to starboard. The two mile buoyed channel then opens up. To the east is the imposing Charles Fort; to the west Fort James. There is a choice of three marinas; Kinsale Yacht Club (tel: 021 772611), Castlepark (tel: 021 774959), and the Trident Hotel (tel: 021 477 2301) which is quite small. Castlepark charges £12 per day for boats up to 30ft, £14 - 30 up to 40ft and £17 - 40 up to 50ft. Corresponding weekly charges are £70, £85 and £105 respectively. For larger boats enquire at the marina. Showers are available for £1. Kinsale Yacht Club charges for an overnight £10 up to 8 metres. Add one pound for each extra metre. In addition the harbour charges dues of £3 for each day for boats up to 11 metres, and £5 for larger boats. If you pay for two days there is no further charge for five days (ie a week in total) when dues kick in again. Visitors can avail of showers (tokens £1), and use the very pleasant bar which serves evening meals.

Kinsale is a beautifully situated town on the slopes of Compass Hill, overlooking the Bandon river. It has had a unique place in the turbulent history of Ireland. Long a busy and strategic port for the British navy, it lost much of its business to Cork Harbour and fell into decline. The traditional black cloaks which its womenfolk wore up until recently, were a direct link to the past. It is only in recent years that it has once again prospered, and is a bustling and vibrant town, that is highly popular with visitors. The timber fronted buildings are unique.

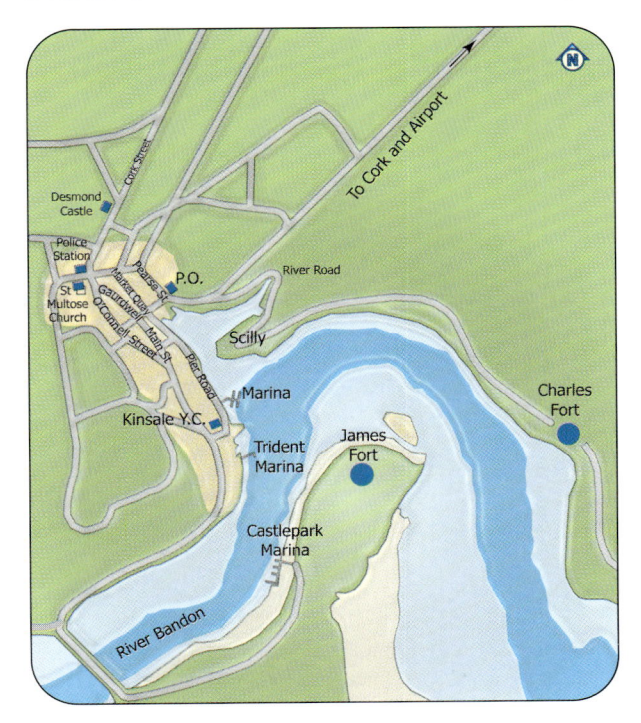

With Anglo-Norman origins, Kinsale became the stronghold of the De Courcy's, whose descendants still live in the area. The year 1601 is pivotal in its history, when the Spaniards under Don Juan d'Aguilla occupied it. The town was besieged by the Lord Deputy Mountjoy. From the north of Ireland an army commanded by the great Irish Earls, O'Donnell and O'Neill, marched south to aid it. But their plans were thwarted by treachery and dissension and Mountjoy prevailed with the Spaniards yielding the town. The defeated Irish chieftains and many followers called 'the wild geese', subsequently left Ireland, in what became known as 'the flight of the earls' (see entry on Rathmullan). But their names surface in the history of mainland Europe where they fought and established themselves, many in the wine trade. In 1689, James II of England sailed from the town to try to regain his throne. He was defeated at the battle of the Boyne and sailed from here into exile in France in 1691. The town was subsequently captured by William of Orange. This episode marked the end of the Irish effort to dislodge the English, and two and a half centuries would pass before an independent Irish State was established.

Worth visiting is the 12th century church of St. Multose, with its curious tower and spire. This church built by the Anglo-Normans has been in constant use as a place of worship right up to the present day. Also worth visiting is the 15th century Desmond Castle which houses the international Museum of Wine and here can be seen the links between the wild geese and many famous wines and cognac, carrying names like Hennessy, Lynch, Barton, Kirwan, MacCarthy and Dillon. It is open daily Mid-June to early October, 1000 to 1800. Entrance is £1.50 for adults and 60p for children, OPAs and groups £1.00. In the Old Courthouse there is a museum with some local and *Lusitania* artefacts.

There are delightful walks to be taken around the town, where fine views of the harbour can be obtained. Charles Fort, about a mile and half from the town centre, is a good example of a star shaped fortress and worth visiting. It was constructed in the late 17th century, and is one of the largest military forts in Ireland. It is open mid-March to October daily from 1000 to 1800; at other times open Saturday and Sunday, 1000 to 1700. Admission is £2.00 for adults, children £1.00, OAPs and groups £1.50 and families £5.00.

In recent years Kinsale has revived its tradition as a source of arts and crafts. A stroll through the winding streets reveals a fine range of locally manufactured pottery, silverware, candles and glassware. Many painters and sculptors have made Kinsale their home.

Restaurants and pubs

Kinsale stages an annual gourmet festival, and the blackboard menus outside the many restaurants are testimony to the culinary reputation of the town. This is just a selection:

Crackpots, Cork Street, some say the best in town, tel: 021 772 847.

Annelies, O'Connell Street, a pleasant bistro, tel: 021 773 074.

Blue Haven, Pearse Street, expensive, tel: 021 772 209.

Paddy Garibaldi's, Lower O'Connell Street (opposite Yacht Club), serves a good range of pizzas and other Italian dishes, tel: 021 774 077.

The White House serves good pub lunches, tel: 021 772 125.

The Spaniard Pub, a pleasant stroll up the hill, on River Road, in Scilly, has good pub food.

Man Friday, Scilly, good old fashioned restaurant, tel: 021 772 260.

Also in Scilly is the popular **Spinnaker Restaurant**, tel: 021 772 098.

Kinsale Gourmet Food and Seafood Bar, beside the church of St. Multose is an excellent spot to have lunch and buy fish and other delicacies.

The **Bulman Pub** and **Bistro** near Charles Fort, is worth the 25 minute walk for good food and drinks and views. Booking advisable at weekends, tel: 021 772131.

Courtmacsherry

Latitude: 51°38'.20N; Longitude: 08°41'.50W;
Charts: Admiralty 2081 2092; Imray C56

Leaving Kinsale Harbour, the Old Head and lighthouse are just west of south. Once around the Head is Courtmacsherry Bay. Courtmacsherry, a lifeboat station, is a pleasant village on the Ardigeen River entered at Wood point. Consult chart for hazards in Courtmacsherry Bay. River entry should not be attempted at low water, enter with at least an hour either side of low water. The channel is marked. There is a small pontoon, where it is possible to lie alongside, at the village with water and power available. There is a grocery and Post Office. For drinks try The Pier House Bar also serving sandwiches. Further up the village is The Courtmacsherry Hotel serving meals, with bar attached, and The Lifeboat Inn and Restaurant. Enquire from locals about walks including a pleasant hike to Wood Point.

Glandore

Latitude: 51°33'.70N; Longitude: 09°07'.20W;
Charts: Admiralty 2092 2424; Imray C56

Having left Courtmacsherry Bay and turning west, past the Seven Heads, Clonakilty Bay is to the north. Around Galley Head, the entrance to Glandore Harbour is to the west. It is entered between Adam's Island to the west and Goat Head to the east. Just over half a mile to the north west is Eve Island. And heed what the locals say – that is to shun Adam and hug Eve! The channel is clearly buoyed and is used by fishing trawlers going up to Union Hall Harbour which lies a mile to the west of Glandore village. There are six visitor mooring buoys located towards the slipway, in the anchoring area. The village looks down on the whole proceedings and is a short row ashore.

The remains of various fortifications and castles can still be seen, some now incorporated into modern buildings. The village of Union Hall, a pleasant walk of just over a mile, has associations with Jonathan Swift, author of *Gulliver's Travels*. He stayed in the village in 1723 to help recover from affairs of the heart, in fact the death of his beloved Vanessa, and was so smitten by its beauty that he wrote a poem in Latin – *Carberiae Rupes (The Rocks of Carberry)* in praise.

Another walk of about a mile and a half will bring you to the Drombeg Stone Circle, which is clearly sign posted. This as the name implies, is a circle of stones dating from about 1000 B.C., making it 3000 years old. The stones are aligned in such a way that on the shortest day of the year December 21st, the winter solstice, at sunrise the sun shines between the fifth and sixth stones onto the fourteenth stone, making it a very early form of calendar. Similar arrangements are to be found in sites such as Newgrange in Co. Meath. Nearby can be seen the foundations of two huts and a cooking pit.

Glandore has two pubs. Here you can sit with your crew, quaffing the drink of your choice and gaze down on your boat securely buoyed, with the beautiful harbour vista and Adam's Island clearly in view, and across the harbour nestling in wooded slopes, Union Hall.

Restaurants:

The Rectory, a short walk, good but on the expensive side, tel: 028 33072.

Marine Hotel, has a pleasant bar and restaurant, tel: 028 33366; **La Scala** a reasonably priced Italian restaurant is next door.

Pubs:

Hayes Bar serves pub grub. **The Glandore Inn** also provides pub grub and in addition has showers which visiting yachtsmen are welcome to use (there is a charge).

The walk to Union Hall, a very pleasant village, which has a Post Office, will let you taste the seafood delights in **Dinty Cullen's**, (also has a restaurant, tel: 028 33373), or **Casey's** bars and there is a good restaurant – **The Bayberry**, tel: 028 33605.

Baltimore

Latitude: 51°29'.30N; Longitude: 09°22'.30W;
Charts: Admiralty 3725 2129; Imray C56

Leaving Glandore and turning south west the entrance to Baltimore Harbour is about twelve miles distant. Off Toe Head lie the dramatic Stags, a group of vertical jagged rocks. There is safe passage in Stag Sound, between the rocks and the head. Then, a couple of miles distant lies the entrance to a highly interesting natural reserve. This is Loch Hyne, nearly landlocked, and well worth a stopover. There is safe anchorage at the entrance to the Lough, at Barloge Creek. The entrance is narrow and negotiable only by dinghy. Care is required as the tide funnels in and out of the narrow entrance at some speed. What makes the Lough unique is the way the tide races in; but because of the depth and configuration of the terrain, ebbs at a slower rate. The result over the years is that fish, including crustaceans, many of them rare and from sub-tropical waters are swept into the Lough but are not carried out on the ebb. There has been a great build-up of species over the years and they are the subject of research by the Marine Biology Department of University College Cork. A display board at the north end of the Lough describes the species found, including purple sea-urchins, starfish, coral and anemones.

Baltimore is a sheltered fishing and boating centre, located to the north east of Baltimore Harbour: entrance is between the lighthouse at Barrack Point to the west on Sherkin Island and Beacon Point to the east, leaving the Loo Rock Buoy to starboard. Atop Beacon Point is a splendid white stone torpedo shape, known locally, with biblical origins as 'Lot's Wife'. Not that she could see much behind, given where she is perched. Once inside the harbour the hazards of the aptly named Lousy Rocks and Wallis Rocks are clearly marked. There is a floating barge against where visiting yachts can tie up with water and power. Charges are per night, £10 up to 35 feet, £12 to 41 feet and £15 above this. Check in at the ferry office on the pier. Showers are available for a £2 charge.

Baltimore or Dún na Sead – The Fort of the Jewels, was a stronghold of the O'Driscolls, the remains of whose castle can still be seen near the pier. The town was raised by Waterford men in 1537 in revenge for the seizing of one of their ships by the O'Driscolls. Later in 1631, pirates from Algiers again razed the town and took many prisoners who were transported to Algiers and sold into slavery. Piracy was rampant in the 17th century. This exploit is commemorated in a poem by Thomas Davis – *The Sack of Baltimore*.

>'The summer sun is falling soft on Carbery's hundred isles –
>Upon that cosy creek there lay the town of Baltimore'

The harbour has a lot of sailing activity; it is a popular cruising spot, and the Baltimore Sailing Club runs regular events including sail training. Visitors are welcome to race in the Baltimore Regatta, in early August.

A short walk brings one to the town centre where there is a Post Office and a variety of pubs and restaurants:

Chez Youen, a restaurant with a good reputation for fish, expensive, tel: 028 20136.

The Mews restaurant, located in an attractive stone building, tel: 028 20390.

The Custom House, also with a good reputation for fish, and has a Bib Gourmand – Michelin award for good food at modest prices, tel: 028 20200.

The Life Boat restaurant serves coffee, tea and light meals.

Baltimore Hotel, (with Casey's Bar), tel: 028 20197.

La Jolie Brise is a cafe, which also serves pizzas, also oysters.

Three pubs provide a convivial outdoor seating area; **Bush's**, **MacCarthys** (also has music sessions) and **The Wheelhouse Bar**. All serve bar food. The aptly named **Algiers Inn** has a pleasant beer garden.

Sherkin Island

Latitude: 51°28'.50N; Longitude: 09°23'.70W;
Charts: Admiralty 2129 3725; Imray C56

Sherkin Island forms the western end of Baltimore Harbour. This island, still well populated, is just over two miles in length, and about a mile across. It is a popular summer holiday retreat. There is a pontoon located beneath Murphy's Hotel, where showers are available. There is also The Jolly Roger Bar. The hotel is clearly visible with its yellow painted walls. Nearby are the reasonably well preserved Franciscan friary ruins. The pontoon has power and water.

The hotel has a bar and attached is the Islander Restaurant. The Cuisin snack bar is located at Silver Strand.

A boat moored to a visitor mooring buoy in Glandore

Schull

Latitude: 51°31'.50N; Longitude: 09°32'.30W;
Charts: Admiralty 2129 2184; Imray C56

Leaving Baltimore passage can be made through Gascanane Sound off Cape Clear Island to Long Island Bay. To the south west lies the famous Fastnet Rock, with its much photographed lighthouse. On the southern side the stump of the original lighthouse, built in metal in 1854, can be seen. It succumbed to the Atlantic weather in 1881 and was replaced by the present granite structure in 1904, with its base secured on lower ground. Every second year the Fastnet Race takes place when yachts leave the south of England, during Cowes Week, race around the Rock and head for home. Not to be outdone, Schull has a very well supported regatta every August, called Calves Week (got it?!). Visitors are welcome to participate.

After passing between Middle Calf Island and East Calf Island, Schull is entered between Long Island (Copper Point) and Mweel Point on Castle Island and thence between the Bull Rock and Coosheen point. Schull village, pier and jetty are located at the northern end of the harbour. There are 12 visitor mooring buoys located about a half mile east of the landing jetty. The jetty can be quite busy with dinghy sailors launching and coming ashore. Repairs can be undertaken at nearby Rossbrin Boatyard. There are frequent ferry services to the nearby islands including Cape Clear.

Schull lying under the slopes of Mount Gabriel, derives its name from the Irish for school – scoil. The monks of nearby Ros Ailithir founded a school here. Today it is a popular holiday and cruising destination. The Schull Community College houses a planetarium; for details of starshows and entry fees telephone 028 28552.

There is a bank and Post Office and a good selection of pubs and restaurants, in the Main Street. **The Courtyard** has a coffee shop, gourmet delicatessen, bar and restaurant. They have music sessions and a beer garden, tel: 028 28390.

La Coquille, a French restaurant, has a good selection of seafood, is open for lunch and dinner, tel: 028 28642.

The Waterside Inn has a restaurant and also serves bar food, tel: 028 28203.

Adele's is a coffee shop and restaurant, tel: 028 28459.

East End Hotel, restaurant and bar with food, patio garden, tel: 028 28101.

The Bunratty Inn serves bar food, tel: 028 28341.

Another coffee shop worth trying is **Cotter's Yard**.

For pubs there is the **Tigin**, **T.J. Newman's**, **The Galley Inn**, **Regan's** and **Hackett's**.

Crookhaven

Latitude: 51°28'.20N'; Longitude: 09°43'.50W;
Charts: Admiralty 2184; Imray C56

Once clear of Schull Harbour, and into Long Island Bay, Crookhaven lies about six miles to the south west. It is a delightful sheltered inlet entered from the east, between Blackhorse rocks and Rock Island, and is just under two miles in length. The village and pier are located about half way up the inlet on the southern shore. Eight visitor mooring buoys have been located opposite the pier, eliminating the problem of poor anchor holding.

Ashore there are a few watering holes in which to unwind in this congenial village, where everyone chats to everyone, and the pace is relaxing.

The **John Dory** restaurant has a good reputation for fish, tel: 028 35035.

For bar food there is **O'Sullivan's** and the **Crookhaven Inn**.

Glengarriff

Latitude: 51°44'.90N; Longitude: 09°32'.30W;
Charts: Admiralty 1838 1840; Imray C56

Leaving Crookhaven and turning south west, the lighthouse of Mizen Head is visible, marking the south westerly extremity of Ireland. Around this lies a series of magnificent bays stretching along the coasts of Cork and Kerry. Here you will find some of Ireland's, indeed possibly Europe's, most breathtaking scenery. (Don't listen to the cynics – they say if you can see the mountains it is going to rain and if you can't – well it is raining!) Once past Dunmanus Bay the great expanse of Bantry Bay opens up. Glengarriff is truly one of the most beautiful corners of this region. It lies at the end of Bantry Bay which extends about 20 miles ENE of Sheeps Head. To reach the visitor mooring buoys, (there are six) located close to the pier, go between Big Point and Gun Point, then leave Garinish and Ship Islands to the west.

Travelling up Bantry Bay, beware of fish farm installations which are clearly visible in daylight. To the east is Whiddy Island, with its massive oil storage tanks visible. The Whiddy Island terminal was built in the early 1970s to accommodate the new supertankers. Oil was transhipped into smaller tankers which would then travel to smaller coastal terminals in Ireland and the UK. The sad remains can now be seen of the unloading jetty. In January 1979, a French tanker, the *Betelgeuse*, was unloading oil, when it caught fire and subsequently exploded. Over 40 ship and shore personnel died and many more were injured. It was a massive conflagration, which burned for days. Today there is still a large amount of oil stored in the tanks, with a limited transhipment taking place.

Glengarriff provides a haven of peace away from those sights and memories. Because of its sheltered position it has luxuriant semi-tropical vegetation. Great masses of fuschia, arbutus and hydrangeas abound. The most famous feature is Garinish Island, or Ilnacullin, reached by a regular ferry service. This Island was acquired by Annan Bryce, architect and landscape artist who, exploiting the mild climate, planted it with a wide range of trees and shrubs imported from many parts of the world, thereby creating a garden of rare beauty. He also added a pagoda and reflecting pool. It was on Garinish Island that George Bernard Shaw wrote one of his best known plays – *St. Joan*. The Island has been donated to the nation, and is open to visitors. There is a landing fee for those visiting; £2.50 for adults, children £1.00, OAPs and groups £1.75 and families £6.00. Opening times are July and August – 0930 to 1830 (Sundays – 1100 to 1900); April, May, June and September 1000 to 1830 (Sundays – 1300 to 1900). If using the ferry it costs £5 for a return journey.

Glengarriff boasts a wide range of pubs and eateries.

Casey's Hotel, has a restaurant and also serves bar food, tel: 027 63072.

Paddy Barry's, is a pub serving food but also has a restaurant.

The Rainbow Restaurant, tel: 027 63440.

Eccles Hotel, tel: 027 63003.

Also **Cafe de Paix Bistro** and **Harringtons**.

Adrigole

Latitude: 51°41'N; Longitude: 09°43'W;
Charts: Admiralty 1840; Imray C56

Adrigole Harbour is located in a small inlet off Bantry Bay, about eight miles south west of Glengarriff. It is entered between Adrigole Head and Drumleave Point. Beyond the slipway which is used by The West Cork Sailing Centre, run by Niall and Gail MacAllister whose building is clearly visible, there are eight visitor moorings. The village of Adrigole is just a short walk from the slipway. Dominating the skyline is the rugged Hungry Hill immortalised in a Daphne du Maurier novel of that name and subsequently filmed. The copper mines, around the fortunes of which story is woven, were in fact located somewhat further afield in Allihies. Clearly visible is Adrigole Cascades falling 200 metres down the mountainside, the highest waterfall in Ireland.

The village has two pubs **Murphy's** and **Thady's**. There is a Post Office, general shop and telephone. The MacAllisters now have a new, purpose-built sailing centre with showers and a launderette.

Lawrence Cove

Latitude: 51°38'.20N; Longitude: 09°49'.20W;
Charts: Admiralty 1840; Imray C56

Bere Island is located on the south westerly end of Bantry Bay, a picturesque and peaceful haven. Lawrence Cove is situated on the northern side of the island across from Castletownbere. The marina is clearly visible from the entrance and there are also four visitor mooring buoys close by. The marina (tel: 027 75044) operated by the developers John and Phil Harrington, has a toilet, shower block and launderette, diesel, water and electricity, and a craft shop. Charges are £1.20 per metre per day, with reductions for longer stays. It is the most congenial in Irish waters. A Harrington welcome is special.

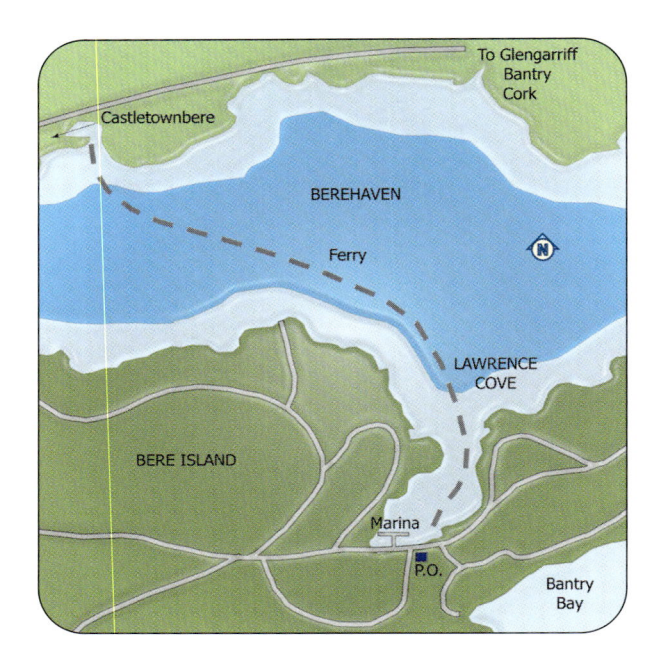

The island, about five miles in length and just over a mile in width, has been witness to many historic happenings all associated with the surrounding sea. On two occasions French fleets entered Bantry Bay, the first time in 1689 to aid James II. In 1796 they came again this time to aid a republican insurrection, but were dispersed by a storm and were forced to return to France. On board was the father of Irish republicanism, Theobald Wolfe Tone.

Bere island subsequently became a major base for the British Atlantic Fleet. It was said by locals that during the First World War you could walk all the way from Castletownbere to Bere Island across the decks of British warships. Although Ireland gained independence in 1922, under the treaty, Britain retained control of her installations at Spike Island in Cork Harbour, Lough Swilly in Co. Donegal and Bere Island. Following the ending of a protracted economic war between Britain and Ireland, Britain ceded the ports in 1938, and the last British military personnel left. The deal negotiated between Neville Chamberlain, British Prime Minister and Eamon de Valera, Irish Taoiseach, is said to have left Winston Churchill characteristically fulminating. Churchill used the ports issue as a bone of contention surrounding Ireland's neutrality in World War II. The island was used as a base by the Irish Army, but they are now withdrawing. The remains of a major fortress complete with moat can be seen on high ground. Two six-inch guns, with their barrels painted in camouflage green, still point seawards. There is a regular ferry service running between Lawrence Cove and Castletownbere, useful for facilitating crew changes and trips to the mainland. There is a general shop and Post Office – Murphy's.

Within a short walk of the marina, there are two restaurants and a pub:

Lawrence Cove House, the 1999 Seafood Restaurant of the Year, tel: 027 75063.

Kitty's Cafe and Restaurant, tel: 027 75996.

O'Sullivan's Pub, where you will be brought up to date on the island's latest gossip.

Castletownbere

Latitude: 51°39'N; Longitude: 09°53'.30W;
Charts: Admiralty 1840; Imray C56

Castletownbere lies to the west of Bere Island and is a major fishing and fish processing port. There are four visitor mooring buoys located to the north east of Dinish Island. The visitor buoys are somewhat distant from the town and a landing location. It is usually possible to overnight at the town pier, but first consult the Harbour Master (tel: 027 70220).

Castletownbere has all the usual facilities – banks, Post Office, supermarkets, hairdressers and there is a good selection of eateries and pubs serving food:

O'Donoghues, very good pub grub.

Jack Patricks a restaurant and butchers shop, so the meat is exceptionally good.

Nikis, tel: 027 70625.

The Old Cottage, tel: 027 70430.

Breens Lobster Bar.

MacCarthy's, a pub with nautical connections.

Dromquinna Pontoon

Latitude: 51°52'N; Longitude: 09°40'W;
Charts: Admiralty 2495; Imray C56

Leaving Bantry Bay, and Co. Cork, passage can be made through the narrow Dursey Sound, over which a cable car runs between the mainland and Dursey Island (cables are about 25 metres above MHWS). Beware of strong tides in this narrow passage, and watch for the Flag Rock. An alternative route is around Dursey Head, between Dursey Island and The Cow Rock. Note the positions of the Calf and Lea Rocks. The Kenmare River is entered between Cod's Head and Lamb's Head. Despite the name, it is not a river but a long wide bay similar to but not as large as its neighbour, Bantry Bay. In the 1700s Lord Lansdowne, a riparian landowner, was responsible for the misnomer in the interest of acquiring for himself extensive fishing rights in the so called 'river'. You are now in the County, or Kingdom as the locals have it, of Kerry. About 15 miles upstream lies Dunkerron natural harbour just south west of Kenmare. In the harbour is Dromquinna Pontoon on which yachts can tie up, but it is prone to silting. It is better to pick up one of the visitor moorings and call the pontoon on 064 42255.

There is water on the pontoon and showers are available. A charge of £6 per day for boats up to 25 feet or slightly more if over that is made. There is a restaurant beside the pontoon serving lunch and dinner. A short walk brings you to Dromquinna Manor Hotel, set in beautiful parkland. The town of Kenmare is a short taxi ride (shore personnel can arrange). The town has all the usual amenities, banks, Post Office, pubs, good restaurant and a launderette.

Sneem

Latitude: 51°48'.70N; Longitude: 09°53'.60W;
Charts: Admiralty 2495; Imray C56

Sneem lies in a beautiful natural harbour off the Kenmare River. Nearby Parknasilla is clearly visible with its luxury hotel in what must be one of the most superb hotel locations in Ireland. There are three visitor mooring buoys located here, but it is about a mile and a half walk to the village of Sneem. Alternatively you can row over to the hotel slip and they will organise a taxi. Sneem is a small town with a bank, Post Office, pubs and restaurants, of which Sacre Coeur is the best.

Derrynane

Latitude: 51°45'.60N; Longitude: 10°08'.80W;
Charts: Admiralty 2495; Imray C56

Cruising yachtsmen describe Derrynane just off the Kenmare river as being one of the most perfect anchorages. However, care must be taken entering. From the south the passage is between Deenish and Moylaun Islands. There are leading lights at the entrance. Beware of entering if there is a south westerly swell. The scenery here is breathtaking and the harbour is framed by a sandy beach. There are three visitor mooring buoys. If lying to an anchor use plenty of chain and lay a kedge. The anchorage is prone to sudden severe gusts through a gap in the surrounding hills.

Derrynane apart from the natural beauty, is synonymous with the name of Daniel O'Connell – 'The Liberator', one of the greatest political figures of 19th century Ireland. He was born in nearby Cahirciveen, but was adopted by an uncle and reared in Derrynane House, which he subsequently inherited. He studied in London and Paris, where the brutality of the French Revolution which he witnessed, turned him against violence as a means to political ends. This was reinforced by what he witnessed of the 1798 rebellion in Ireland. He entered politics and was elected MP for Clare.

The achievement for which he is best known, was the granting of legal rights to emancipation of Catholics in 1829. These had been denied Catholics under the notorious penal laws. His next task was to seek the repeal of the Act of Union of 1801, which subsumed Ireland into the United Kingdom. He had enormous organisational and oratorial powers and the British, from their imperial standpoint, were very wary of the prospect of an independent Ireland. At one stage he was even arrested and imprisoned for conspiracy. He organised gigantic public meetings to advance the repeal cause, one of which attracted three quarters of a million people to The Hill of Tara, in Co. Meath.

In failing health, he set out for Rome in 1847, hoping to end his days there. He never reached his goal and died in Genoa. But his heart is buried in Rome, whilst his body was returned to Ireland and is buried in Glasnevin Cemetery in Dublin; a 165 foot round tower, marks his grave. Another monument commemorates him in the main thoroughfare of Dublin called after him – O'Connell Street. Here he stands surrounded by angelic looking ladies, although the wags say that in his lifetime, some of the ladies he knew might not have been so angelic. And these statuesque bronze ladies still carry the bullet holes they received during the Easter Rising in 1916.

Derrynane House, situated in over 200 acres of parkland on this scenic coastline is open to the public, and has a display of memorabilia covering Daniel O'Connell's life and career. From May to September it is open from 0900 to 1800

(Sundays – 1100 to 1900). Admission is £2.00 for adults, £1.00 for children, OAPs £1.50 and groups and families £5.00.

Derrynane Hotel, tel: 066 947 5136.

Keating's Pub is close by and offers showers.

Valentia Island

Leaving Derrynane and heading northwest Bolus Head and Puffin Island are rounded to enter Portmagee Channel between Valentia Island and the mainland, Bray Head and Reencaheragh Point. The island is seven miles in length, two miles wide and connected to the mainland by a road bridge. The Spanish sounding name is derived from the Irish for a nearby sound, Beal Inse. Valentia is probably best known as the terminal for the first transatlantic telephone cable, completed after many attempts in 1866. The cable came ashore at Knightstown on the eastern end of the island. Valentia has always been a popular holiday destination, a real chance to get away from it all. Sandstone slate was mined and exported all over the world until the late 19th century and the quarry is now a visitor amenity area. Footprints of amphibian dinosaurs have been discovered. Visitors come for the deep sea fishing and it is the base for boat excursions to the dramatic pinnacle-like Skellig rocks.

There are two Skelligs, the larger is Michael; the other is known as Little Skellig. There is an interpretative centre, across the bridge from Portmagee, which traces the origins and history of the Skelligs, and is a must for visitors. With the establishment of this excellent centre, the archaeological authorities are discouraging visitors from landing on Skellig Michael, in an effort to preserve the site. In any case this can only be attempted when there is no Atlantic swell. In spring and summer, Little Skellig can appear to be shrouded in a white mist. It is no mist, but rather the tens of thousands of swirling gannets which nest there. This is the largest gannetry in the world.

On Skellig Michael (which ranks with the Pyramids of Egypt, as a UNESCO World Heritage Site), are some of the earliest Christian monastic remains extant. One might well ask why the monks chose such an inaccessible place. Kenneth Clark in his book *Civilisation* provides the clues. The Church in the 6th and 7th centuries was spilt and riven by controversy. Islam had been established and was flourishing, effectively blocking the expansion of Christianity to the east. So the monks looked west and to isolated spots, like Skellig Michael and the Island of Iona in Scotland, where religion could survive and reach out in more favourable times.

From the landing jetty, a stairway built by the monks and still useable, leads to the very well preserved remains of the settlement. Established in the 6th century, it continued in use until the 12th century. The condition of the remains is a testament to the strength of their construction, given their exposure to Atlantic weather. There are six beehive-shaped stone huts, and two oratories. Here the monks prayed and meditated in these terribly remote conditions. They recorded the gospels in illuminated form onto animal skins, ensuring their preservation. They traded fish and bird feathers for their daily needs. From tiny gardens which they built with soil and seaweed garnered from wherever possible and sheltered by stones, they grew vegetables. Five water storage wells into which rain water was channelled are still intact.

There are extensive monastic and other remains on Valentia, and plenty of scope for walking.

Portmagee

Latitude: 51°53'.30N; Longitude: 10°22'.50W;
Charts; Admiralty 2125; Imray C56

The village of Portmagee is on the mainland, at the road bridge, a couple of miles from the channel entrance. There are six mooring buoys located across from the village near the Interpretative Centre. Unfortunately, the moorings are some distance, about a third of a mile, from the slipway.

Portmagee has a good selection of restaurants and pubs:

The **Fisherman's Bar** serves pub food and incorporates the **Skellig Restaurant**.

The **Bridge Bar** serving good pub food.

The **Mooring's** restaurant, tel: 066 947 7108.

The **Skellig Mist** coffee shop.

Knightstown

Latitude: 51°55'.70N; Longitude: 10°17'.30W;
Latitude: 51°55'.50N Longitude: 10°17'W;
Charts: Admiralty 2125; Imray C56

The centre span of the bridge can be opened on request by ringing 066 77174. The Portmagee channel then takes one to Knightstown at the eastern end of the island, home of the Valentia Island Lifeboat. The alternative is to round the north west side of the island. There are two sets of six visitor mooring buoys, one east and one west of the harbour. Knightstown is a colourful village with an imposing clock tower. It once was home to the many telegraphers and clerks who manned the cable station, which closed in 1965. There is a frequent ferry service to Caherciveen, a good size town, where a marina is under construction.

Restaurants and bars:

Moriarity's Restaurant, for steaks and seafood, tel: 066 9476204; **Boston's Bar** which incorporates **The Tailor's Loft Restaurant**; **The Islander Restaurant**; **Royal Pier Bar**.

Kells

Latitude: 52°01'.60N; Longitude: 10°06'.30W;
Charts: Admiralty 2789; Imray C56

From Valentia and around Doulus Head, the magnificent expanse of Dingle Bay comes into view. Kells is a pleasant inlet on the southern side of Dingle Bay. There are four visitor mooring buoys, some of which can unfortunately be occupied by local lobster boats. The inlet is some distance from the village of Kells, but The Sea View Farmhouse on the shore, offers simple meals and will arrange showers.

Dingle

Latitude: 52°08'.14N; Longitude: 10°15'.48W;
Charts: Admiralty 2789 2790; Imray C56

Dingle lies at the south western end of the Dingle Peninsula, the most northerly of the three peninsulas that jut out from the Kerry coast into the Atlantic. The entrance is well buoyed. It is a busy fishing port with much trawler traffic. The fishing harbour is to the east, and the marina (tel: 066 51629), which is located to the west, has 80 berths, 20 reserved for visitors. The visitor overnight rate is £1.10 per metre. All the usual facilities are located alongside; showers, water and fuel.

Dingle derives its name from the Irish for fort, daingean, which refers to the Celtic fortress which once stood here. To the north is Mount Brandon, the fourth highest mountain in Ireland, which is traversed by the famous Connor Pass. Dingle is a pleasant market town, whose narrow streets are very busy with visitors during summer. It has all the usual amenities; banks, Post Office (Main Street), hotels, launderette and supermarkets. Dingle Aquarium is located on the seafront. There are regular boat trips to see Fungie, a bottlenose dolphin, who has made the harbour his home since 1983.

Dingle hit the world stage in the early 1970s when David Lean chose the town and environs for the making of his film *Ryan's Daughter*. Stars such as Robert Mitchum, Trevor Howard and John Mills were familiar figures on the streets. It was well chosen as you need never go thirsty in Dingle. It is reputed to have no less than 52 pubs! There are several on the waterfront, these include:

The Marina Inn, which serves food all day, as does the **Máire de Barra**.

Murphy's is a combined pub and restaurant.

An Drocead Beag in Main Street, is a pub with music.

Dick Mack's in Green Street combines a shoe shop and pub, and there are regular singalongs in the evening around the piano.

Restaurants:

Benner's Hotel, Main Street, houses the **Atlantic Restaurant**, and **Mrs. Benner's Bar** also serving food, tel: 066 915 1638.

For seafood try the **Armada**, tel: 066 915 1505, **Dannos**, tel: 066 915 1855 and **Longs**, tel: 066 915 1231.

Restaurant Beginish in Green Street, somewhat pricey, but excellent, tel: 066 915 1588.

Doyle's in John Street famous for its seafood, expensive, tel: 066 915 1174.

The Half Door also in Green street, another seafood location.

The Chart House, The Mall, managed by the McCarthys and much praised, tel: 066 915 2255.

Ventry

Latitude: 52°07'.70N; Longitude: 10°21'.40W;
Latitude: 52°07'.10N Longitude: 10°21'.90W;
Charts: Admiralty 2789 2790; Imray C56

Around Reenbeg point lies Ventry, or Ceann Trá, translated from the Irish as 'head of the strand'. It is west of Dingle and is set in its own picturesque bay. This part of the Dingle Peninsula is a gaeltacht or Irish-speaking area. When travelling around this area, this becomes clear as the road signs and many of the shop fronts are in Irish. Ventry is a small hamlet in an area of great historical and archaeological interest. This part of the Dingle Peninsula is rich in early Celtic remains. There are many ringforts, stone crosses and pillarstones, clochans or stone beehive huts. Many stones are carved with Ogham, ancient Celtic writing. Probably the most famous artefact is the Gallarus Oratory or Church (see entry on Smerwick).

There are six visitor mooring buoys, in two sets of three, located just to the east of the slipway. In the village are The Ventry Inn, a pub which offers meals in the summer, and the Skipper Restaurant offering reasonably priced meals.

Smerwick

Latitude: 52°11'N; Longitude: 10°22'.50W;
Charts: Admiralty 2789; Imray C56

Leaving Ventry Harbour and rounding Parkmore Point, Slea Head under towering Mount Eagle, is to the west with the Blasket Islands beyond. There are four islands. The Great Blasket is the largest and was inhabited up until the 1950s when the islanders were moved to the mainland. Storms frequently left them isolated for days and even weeks, short of food and essential supplies. Shipwrecks could be a blessing with cargoes washed up. It has the extraordinary distinction that out of just over one hundred inhabitants, three became the authors of best selling accounts of growing up and living there, eking out a precarious existence on the island. These were Peig Sayers who wrote *Peig*, Tomás O Crohán *The Islandman* and Maurice O'Sullivan *Twenty Years a Growing*. Their efforts were inspired by amongst others, the Oxford scholar, Robin Flower who was a regular visitor to the island, and who learned the Irish language. His own account of life on the island is entitled *Western Island*. These books give an extraordinary insight into people who survived in most difficult circumstances. Many other scholars and folklorists followed Flower to the island, from Norway, Germany and England.

Another Blasket island, Inisvicklane, was purchased by the former Taoiseach or Prime Minister of Ireland, Charles Haughey. He maintains a holiday home there and is a regular visitor.

Smerwick is a small harbour on the northern side of the Dingle Peninsula. It adjoins the village of Ballydavid. Nearby traces of a famous fort, Dun An Óir or The Golden Fort, can be seen. The fort was constructed in 1580 by a party of Spanish and Italians. The fortification was subsequently bombarded by English land and seaforces under the command of Lord Deputy Grey and Admiral Winter. The fort capitulated. The officers were spared but 600 foreign soldiers and 17 Irish were slaughtered. Several clergy including Oliver Plunkett had been handed over before the bombardment, but as was customary in Elizabethan times they were subsequently tortured and hanged. The Irish memory is long, and locals still recount those awful events, as if they happened in recent years. Indeed the Smerwick massacre has often been used to stir nationalist feelings. To happier thoughts however as Smerwick Harbour is surrounded by beautiful sandy beaches.

A mile or so from Smerwick, is the famous Gallarus Oratory, which is about 1200 years old. It is a perfect specimen of a corbel-roofed dry masonry structure. It measures about seven by five metres and is five metres in height. It has survived wars and weather and is in a remarkable state of preservation.

There are four visitor mooring buoys easily visible, close to the pier. The village has a pub – Tig Beaglaoic. Tabhairne Ui Choncuir (O'Connor's Tavern) has a bar and a restaurant serving simple meals. Their motto is Ceol Bia agus Deoc (Music, Food and Drink). Within walking distance is Ballydavid Post Office and a small grocery shop.

Fenit

Latitude: 52°16'.20N; Longitude: 09°51'.60W;
Charts: Admiralty 2254 2739; Imray C56

From Smerwick, having rounded Brandon Head and the passed Magharee Islands, the broad expanse of Tralee Bay opens up. Follow pilot instructions, picking up the Little Samphire Lighthouse, west of the harbour. Fenit is the port for the town of Tralee, which is about eight miles distant at the head of the bay. Tralee is the administrative centre for the County of Kerry. St. Brendan, The Navigator, was born here in 483. Local tradition has it that a small beehive structure on nearby Mount Brandon was his dwelling, where he practised self denial. He left his abode and embarked 'In a wicker boat with ox-skins covered oe'r', to seek the land of promise of the saints'. Popular belief has it that he sailed until he reached the continent of America. The story of his voyages is told in a medieval publication *Navigationis Brendani*. These legends were put to the test in 1976 by Tim Severin, who having studied what information was available, constructed a boat made only with materials which would have been available to St. Brendan. His account of building the boat and his voyaging in the saint's footsteps is recounted in the *The Brendan Voyage* which became a best seller.

It was in Fenit that it was hoped to land a cargo of arms for use in the 1916 rebellion. On April 20 1916, the German steamship *Libau*, posing as another ship *The Aud* arrived off Fenit, making contact with a German submarine U19 in the area. However *The Aud* had been shadowed and was then intercepted by British naval ships and escorted to Cork. On board the U19, commanded by Raimund Weisbach, was Sir Roger Casement, a prominent British diplomat, who had achieved distinction for his part in uncovering humanitarian abuse in the rubber plantations of the Belgian Congo and in South America. Casement became a champion of the Irish cause, went to Germany and arranged for arms to be shipped for use in the planned uprising. Despite knowing that the arms importation plan had failed, he insisted on being put ashore, at Bannow Strand on nearby Ballyheigue Bay. He was arrested, tried and hanged as a traitor. The crew of *The Aud* scuttled her off the entrance to Cork Harbour near the Daunt Rock. An interesting footnote to this episode is that Casement had originally embarked on the U20, the submarine which sank the *Lusitania* off the Old Head of Kinsale in 1915, but transferred to the U19, after the U20 encountered technical problems. Weisbach lived to attend the 50th anniversary commemoration of the 1916 rebellion in Dublin.

Fenit has a marina (tel: 066 713 6231) with 110 berths, 30 reserved for visitors and listens on channels 37 and 80. Visitors should report to the Harbour Office (0900 to 2100 June to September). If closed, report to the West End Bar in the village. A smart card will be issued for access to the marina, showers and telephone. Each berth has 240v electricity operated by £1 card tokens obtainable from the office. There is a laundry in the marina services building (washing and drying £2.50 each). Refuse can be deposited in containers at the blue gates adjacent to the marina building. Rates are £1.20 per metre (£12 min.) for overnight stays and £2 to £7 per metre, per week, depending on season, highest rates apply in July and August. A grocery shop (Michael Parkers) is located in the village.

Visitors are welcome to compete in races organised by Tralee Sailing Club and to visit their fine clubhouse overlooking the harbour. At the end of the pier is Fenit Sea World, an aquarium displaying a wide and fascinating range of Atlantic species, including sharks.

Restaurants and Bars:

West End Bar serves food, tel: 066 713 6246.

Godley's Hotel, tel: 066 713 6108.

The Lighthouse Hotel, tel: 066 713 6444.

The Tankard Restaurant, good for seafood, is in Kilfenora about three miles away, tel: 066 713 6164.

Refer to map on page 98.

Carrigaholt

Latitude: 52°36'.20N; Longitude: 09°41'.70W;
Charts: Admiralty 1547 2254; Imray C55

From Fenit there are few anchorages or shelters until the Estuary of the Shannon is entered, between Kerry Head and Loop Head. The estuary provides a scenic and picturesque cruising area. It is also the habitat for a large number of dolphins. Keep an eye out for these delightful creatures. The Shannon is the largest river in Ireland or Great Britain and the estuary is nearly 50 miles in length from the mouth to the city of Limerick. From Co. Cavan, where it rises, the Shannon flows over 100 miles through three lakes, Loughs Allen, Ree and Derg to Limerick. Through canal interconnections, it is possible to traverse Ireland to Dublin, Waterford and into the Lough Erne system in Northern Ireland. The river and lake levels are partly controlled by The Electricity Supply Board, ensuring a constant flow to Ireland's first major generating station, the hydro station at Ardnacrusha near Limerick. It opened in 1928 and still delivers power to the national grid.

Eight visitor mooring buoys have been placed at Carrigaholt, which is about ten miles east of Loop Head. Ships of the Spanish Armada sheltered here in 1588. On the nearby headland are the ruins of a MacMahon castle. Carrigaholt is a small village, with a lot of charm. There is a Post Office and grocery shop and choice of pubs and eateries:

The Long Dock Bar and Restaurant, serves excellent seafood, including lobster with tables outside for alfresco dining, tel: 065 905 8106.

Rahona Lodge, tel: 065 905 8196.

Pubs include **The Bow Way**, **Lanty's**, **The Anchor Bar**, **Keanes**, **Morriseys** and **Carmodys**.

Kilrush

Latitude: 52°37'.90N; Longitude: 09°29'.70W;
Admiralty Charts: 1547 1819; Imray C55

About seven miles from Carrigaholt lies Kilrush also on the northern side of the lordly Shannon Estuary. For many years a thriving commercial port, it fell into decline and the creek silted up. But in the early 1990s a major project was undertaken – this was the construction of Kilrush Creek Marina, and associated sea lock. This ensures a constant water level in the marina, and allows boats to enter from the estuary on any tide.

The entrance to the lock is clearly buoyed. Prior notice should be given to the marina (Channel 80) before arriving at the lock. The marina (tel: 065 9052072) has 120 berths. Visitors are charged £1 (minimum charge £10) for a 24-hour stay, £6 per week, and £15 per month, all prices per metre. There is fresh water and power on the pontoons. Showers, toilets, launderette and telephone are located in the marina building. There is a fully equipped boatyard in the basin, with a 40-ton hoist, for all boat maintenance and repairs. There is also a well stocked chandlery, Glynn Marine Supplies Ltd.

Kilrush Creek Marina is located at the end of the wide and imposing Francis Street. The town has a pleasant Market Square and has a range of shops, pubs and restaurants, banks and Post Office. However, it is about a mile and a half offshore where the most historical interest in the area lies. This is Scattery Island. There is a slipway but it is recommended that you anchor close to shore and land by dinghy. Alternatively, there are regular boat trips from Kilrush. On the island are the remains of a monastic settlement, founded in the 6th century by St. Senan. The remains of six churches and a round tower can be seen. The island suffered from Viking pillaging, but fortunately a great deal remains. Close to the marina is the Scattery Island Visitor Centre which houses an exhibition of the natural and cultural history of the island. It is open from mid-June to mid-September from 1030 to 1830 daily; admission is free of charge. In the Town Hall there is the Heritage Exhibition – 'Kilrush in Landlord Times'.

Kilrush is a centre for sea anglers who head for the Atlantic waters off Loop Head for the 'big ones', including sharks.

Restaurants and Pubs:

There are three pubs in Henry Street which serve food. **The Haven Arms** (tel: 065 9051267) is a favourite stop for sailors. It serves food all day at the bar and in the **Ships Restaurant**. Also serving food are **Kelly's** and **The Colleen Bawn**. **Crotty's** in the Square serves food and has music sessions. Another musical pub is **The Percy French** in Moore Street. **The Quayside Restaurant** in Frances Street serves breakfast, lunch, snacks and you can also taste their Irish baking.

Refer to map on page 99.

Labasheeda

Latitude: 52°36'.90N; Longitude: 09°14'.20W;
Charts: Admiralty 1548; Imray C55

Travelling east from Kilrush, two electricity generating stations come into view; Money Point to the north and Tarbert to the south. There is a regular passenger and car ferry service from Killimer to Tarbert, saving many hours of driving between Counties Clare and Kerry. About 14 miles passage from Kilrush lies Labasheeda Bay, where there are three visitor mooring buoys. There is a quay at the small village of Labasheeda, where thirsty sailors can relax in Casey's bar, which has a grocery attached. There is also a Post Office and public telephone.

Foynes

Latitude: 52°37'N; Longitude: 09°9'.50W;
Charts: Admiralty 1549; Imray C55

About half way up the Shannon estuary, lies Foynes. The estuary is buoyed for commercial traffic going to and from the ports of Limerick and Foynes. The Foynes Yacht Club has installed a pontoon and visitors are welcome to tie alongside. The clubhouse, with bar and showers, is open mainly at weekends.

Foynes is probably best remembered for the fact that it was a flying boat base from the late 1930s until after World War II. Then nearby Rineanna (renamed Shannon International Airport) became established as the last refuelling point for propeller aircraft crossing the Atlantic – today it is a busy domestic and international airport. Many people passed through Foynes going to and from the Americas. Passengers generally stayed overnight while the planes were serviced and refuelled. During World War II many sober suited gentlemen, embarked and disembarked. The Irish Government whilst ostensibly neutral, allowed passenger aircraft to land and refuel throughout the war. Little did the locals know that many of the suited gents had their uniforms in their cases, and were in fact high ranking allied military personnel. The fairway where the flying boats landed and took off is between Foynes Island and the mainland. There is a museum in the village with interesting flying boat memorabilia, and visible on the hill is the airport building which doubled as a hotel for the transiting passengers. A flying boat captain who revisited Foynes and who purchased one of the aircraft and flew in and out of Lough Derg, was the husband of the film star, Maureen O' Hara.

There are several bars and eateries along the Main Street:

Foynes Inn, **Shannon House**, pub and restaurant, tel: 069 65138, **Yankee Clipper Bar** (named after a flying boat), **Ebzery's Cafe**.

Kilronan

Latitude: 53°07'N; Longitude: 09°40'W;
Charts: Admiralty 2173 3339; Imray C55

Leaving the Shannon Estuary and turning north, the Aran Islands come into view, while to the east are the mighty Cliffs of Moher. Further north the mainland with its scattering of grey limestone appears drab. But that drabness conceals a remarkable botanical secret. For in the crevices between the rocks, seedlings carried by winds from as far away as the Swiss Alps have over time fallen to earth and taken root. There, sheltered from the winds and nourished by the temperate Gulf Stream warmed climate, they have flourished. Rare orchids can even be seen. The area known as The Burren is strictly controlled to stop illegal removal of plants and rocks to preserve this unique botanical occurrence.

The Aran Islands nestle together in Galway Bay, stretching in a north westerly direction from the coast of Co. Clare. First is Inisheer or East Island, the smallest of the three. A couple of miles across Foul Sound is Inishmaan, Middle Island; then another couple of miles across Gregory Sound there is Inishmore or The Big Island, about eight miles in length and two miles wide with just under 1000 inhabitants. Largest town and ferry port is Kilronan, whose sheltered harbour is at the southern end of the island and there are eight visitor buoys.

The Aran Islands are renowned for the hardiness of their inhabitants who have traditionally earned a living from the meagre soil. Stones were cleared and used to build sheltering walls to protect crops and livestock from the Atlantic storms. Seaweed and sand add volume and nutrition to the soil. The Aran islanders have always been great fisherfolk putting to sea in the all weathers in the traditional 'currach'. These boats are about six to seven metres in length with interlaced timber frames on which cattle hides were stretched. The hides were latterly replaced by tarred canvas. Normally propelled by oars, a mast and sail can be raised, to take advantage of favourable winds. Currachs have been used to transport everything, from animals to people to barrels of Guinness. Many are still in use today. The Irish language is still the language of the islanders, although English is widely spoken and many signs and shop fronts display English.

The Aran Islands took centre stage in the 1930s when the American director, Robert Flaherty, made the film *Man of Aran*. An enduring classic of its genre, it portrayed the day-to-day lives of these hardy people, with local people playing all the parts. The playright John Millington Synge, whose best known work is *The Playboy of the Western World*, based his play *Riders to the Sea*, on Inishmaan. Another famous literary character, the writer Liam O'Flaherty, perhaps best known for his short stories, was born on Inishmore.

The traditional dress of the women was a red woollen skirt and crocheted shawl. The men traditionally wore báinín or white woollen woven sleeveless jackets with a crios or multicoloured belt and shoes of cowhide on their feet. The Aran sweater has become a very popular garment internationally, the wool providing great comfort and warmth. Families traditionally had their own particular stitches by which their work could be recognised. The Aran sweater is widely available on the islands and elsewhere and still a great favourite with both sexes.

Alas today the traditional dress is all but gone. The islands are served by regular air services and high speed sea ferries. The thatched cottages have been replaced by modern slate roofed bungalows. And so progress has levied its toll on traditional mores and ways.

But the islands are of great interest because of their fortifications and monastic remains. Aranmore in particular was a major religious settlement, founded by St. Enda in the 5th century, where many of the fortifications are still standing. A pleasant mile and a half walk from Kilronan brings one to the highest point of the island. This is Dun Eochall, with a circular Bronze Age fort. Nearby is a disused lighthouse. Close by is Teampall Chiarain, St. Kieran's monastery with its stone remains carved with crosses.

About six miles from Kilronan is Dun Aengus which has been described as the most spectacular prehistoric remains in Europe. The fort covering about five hectares consists of three concentric enclosures. The middle wall is covered by a remarkable abatis or chevaux de frise of jagged stone uprights, designed to foil and even impale invaders. From the fort it is a sheer drop of 100 metres to the ocean below. There is a visitor and exhibition centre about 900 metres from the fort. It is open from March to October 1000 to 1800. Admission is £1.00 for adults, 70p OAPs, children 40p and families and groups £3.00. At other times of the year it is open 1130 to 1530. The terrain here is quite uneven and care must be taken.

Across the island is Clochan na Carraige, an ancient beehive stone dwelling. Also in this area are what are called the Seven Churches, in fact there are two churches, other remains probably being domestic dwellings. Finally, nearer to Kilronan is Dun Dubhcathair or Black Fort, another spectacular cliff top fortification with the remains of a chevaux de frise.

A popular way to tour the island is by bicycle and there are several hire shops in Kilronan, which has a Post Office, bank, grocery shop, pubs and eateries:

Dun Aonghasa and **Aran Fisherman Restaurant and Bar**, serving locally caught seafood, tel: 099 61104.

The Bay Cafe has a seafood restaurant and wine bar, tel: 099 61260.

Joe Watty's Bar serves pub grub, **The American Teach Tabhairne** (Bar), serves sandwiches/snacks, **Lucky Star Bar**.

Struthan

Latitude: 53°16'.30N; Longitude: 09°34'.50W;
Charts: Admiralty 2173 3339; Imray C55

Leaving Kilronan, Cashla Bay is in a north easterly direction across Galway Bay. It is sheltered in all weathers and there are eight mooring buoys at Struthan west of Rossaveal. State of the art high speed ferries leave Rossaveal to whisk passengers to the Aran Islands. Nearby is the airport which serves the Aran Islands. This area is the start of that part of Co. Galway called Connemara, stretching to the west and containing some of the most spectacular scenery in Ireland. Great artists including Paul Henry and Muiris MacGonigal were inspired by the landscape and painters still come to record the natural beauty.

A walk of about a mile brings you to the village of Carraroe, and the Connemara Gaeltacht, which has two hotels, Ostán and Hotel Doilin. There are three pubs; An Tig Taileur (The Tailor's House), An Cistin (the Kitchen) and An Realt (The Star).

Maumeen

Latitude: 53°17'N; Longitude: 09°38'.50W;
Charts: Admiralty 2173 3339; Imray C55

Leaving Cashla Bay and turning west Maumeen is reached by way of Greatman's Bay — consult the pilot for hazards. There are four visitor mooring buoys here. The pier was used until recent times for hookers delivering turf and supplies (see entry on Roundstone for details of these boats). The village of Lettermore is about a mile distant and thirst can be quenched in O'Toole's Pub.

Kilkieran

Latitude: 53°19'.30N; Longitude: 09°43'.40W;
Charts: Admiralty 2096 2173; Imray C54

Leaving Greatman's Bay and turning west, beware of English Rock. Golam Head, with its conspicuous tower, is about four miles distant and marks the easterly side of the entrance to Kilkieran Bay. Kilkieran, a small fishing port lies four miles north of Golam Head. There are twelve visitor mooring buoys located to the east of the pier; further east still, is Lettermore Island.

The factory located in the harbour is operated by Arramara Teoranta and processes seaweed. To the uninitiated this might appear a questionable activity. However, seaweed contains many important mineral elements including iodine, potassium and manganese. But the main uses are in the extraction of alginates for use as a fertiliser especially for enhancing sportsground grasses. Other uses include organic food supplements for livestock, horses and ponies in training and for broodmares. Alginates are also used in a wide range of products which require gelling, thickening or emulsifying. Examples include ice cream, mayonnaise and cosmetics. But most importantly of all the head on your pint of beer or lager, keeps its froth longer, because of – yes – alginate gels! The harvesting of the seaweed is a physical and laborious process, but provides well-paid seasonal employment for local boatmen and their crews.

Kilkieran is a small village with a Post Office and two pubs; **McDonnacdha's** who also have a grocery and butcher and **Peter J. Conroy's** pub which serves sandwiches and has a grocery shop attached.

Roundstone

Latitude: 53°23'.40N; Longitude: 09°54'.60W;
Charts: Admiralty 2173 2709; Imray C54

Leaving Kilkieran Bay passage should be made to the North Sound. When turning west, keep to seaward of the Skerd Rocks. There are numerous small islands and rocks off this part of the west coast and there are few marks, so planning and care are needed. Entry to Roundstone Bay is via Big Sound, lining up with the lighthouse on Inishnee Point. There are four visitor mooring buoys one mile north of Inishnee Point, but they are about a half mile east of the village and harbour.

Roundstone harbour and village were designed and constructed in the early 19th century by Alexander Nimmo, a distinguished Scottish engineer. It has long been a favourite holiday location and haven for painters and botanists, taking in the Connemara landscape and mountains – the Twelve Pins or Bens. The area is renowned for the bracing sea air and white coral beaches. Overlooking the village is Errisbeg Mountain, from the slopes of which superb views can be had of the myriad islands and bays. Roundstone Harbour is very sheltered, but lacks sufficient water at LWN. There is good holding ground for anchoring outside the harbour a little to the south of the north pier.

The 'Galway Hooker' can be seen in this locality and in many harbours on the west coast. These are ketches carrying a main and top sail and two foresails on a bowsprit. They also carry a mizzen. These boats also known by their Gaelic name – pucan, are distinguished by their black hulls and dark red sails. They can be up to 15 metres in length. They were the cargo workhorses of the west coast of Ireland carrying turf, livestock and passengers in the days before road transport developed. They have enjoyed a considerable revival in recent times, and the annual Cruinniú na mBád or gathering of the boats in August at Kinvarra, a fishing port south of Galway City is a major event for them.

Roundstone has a good range of restaurants and pubs, serving locally caught seafood including its famous lobsters:

Restaurant Beola, tel: 095 35871.

Balcony Restaurant.

Roundstone House Hotel, Vaughan's Restaurant, tel: 095 35864.

Eldon's Hotel, tel: 095 35933.

O'Dowds pub and restaurant serves pub grub, famous for its seafood, tel: 095 35809.

Ryan's bar also serves seafood.

The Coffee Dock serves teas and coffees and light meals.

Other bars include, **Kings**, **Connollys** and **The Matchmaker** which has music.

Clifden

Latitude: 53°29'.20N; Longitude: 10°03'.40W;
Charts: Admiralty 1820 2708; Imray C54

From Roundstone the journey takes you around Slyne Head passing the Errislannan Peninsula and into Clifden Bay. Keep about a third of a mile south of the Carrickrana Beacon. There are eight visitor mooring buoys located about two miles away on the northern side of the Errislannan Peninsula. The slipway of Clifden Boat Club and the Inshore Lifeboat Station are about a half mile away to the east. There is good holding ground for anchoring opposite the club.

Clifden is the capital of Connemara, and has very interesting links to the past. It was from nearby Derrygimla that Marconi transmitted the first transatlantic wireless message in October 1907. There was a major transmitting station here, with its own turf-fired electricity generating plant, employing over 70 people until technological changes led to its closure in the 1920s. Little now remains of the installations.

And it was the transmitting masts of this station that guided Captain John Alcock and Lieutenant Arthur Whitten Brown to a boggy landing there on the morning of Sunday June 15 1919. While the inhabitants were in church, these intrepid fliers completed the first transatlantic flight. They had left St. John's Newfoundland sixteen hours earlier in their Vickers Vimy biplane, intending to fetch Galway Bay and find suitable landing there; making the Irish coast so close was a superb navigational feat for its time. They had a smooth landing, but just as they came to a stop the nose wheel went into a bog and tipped the plane up. The fliers were unscathed, just saying that they needed a bath and a shave. The aircraft suffered slight damage to its undercarriage. A memorial in the form of an aircraft tail which points to the exact landing spot can be seen about a mile from the town, on Ballinaboy Hill, off the Ballyconneely Road. The memorial site also affords a magnificent panorama of Clifden and its bay, Ardbear Bay and the Errislannan Peninsula.

As with so many parts of Connemara, sandy beaches abound, many with glistening white coral. Clifden has banks, a Post Office and a wide range of hotels, restaurants and pubs.

Alcock and Brown Hotel, Restaurant and Bar, tel: 095 21206.

Station House Hotel, Restaurant and Bar which also has a leisure centre with a swimming pool call 1850 377000.

O'Grady's Seafood Restaurant, Market Street, famous for its fish, tel: 095 21450.

Clifden Bay Hotel.

Mannion's, pub with bar food.

Guy's, a popular bar.

EJ King's, The Square, good pub grub, music sessions.

Mitchell's, Market Street, medium priced restaurant, good seafood, tel: 095 21867.

Destry's, medium priced restaurant, tel: 095 21722.

Fogartys Restaurant, wide menu including seafood, tel: 095 21427.

Leenane

Latitude: 53°36'N; Longitude: 09°43'W;
Charts: Admiralty 2420 2706; Imray C54

Leaving Clifden there are many islands to negotiate. In the distance is Inisbofin, and further to the north, Inisturk and Clare islands. Leenane lies at the end of the seven mile long Killary Harbour. This natural harbour is in fact a fiord, with very deep water. It was said that the entire British Atlantic fleet could anchor in Killary Harbour and leave space for as many ships again. There are extensive fish farming installations and extreme care is required. On either side the gorse and heather clad mountains slope down to the waters edge. Again, superlatives are needed to describe the scenery here especially on a clear sunny day. Eight visitor mooring buoys are located about a mile from the village which has a Post Office and grocery shop. The village featured in the film adaptation of the play by Kerry writer, John B. Keane, *The Field*. A restaurant in the village commemorates this.

Leenane Hotel and Restaurant, tel: 095 42249, **The Field Restaurant**, tel: 095 42271, **The Village Grill**, **Gaynor's Bar** and **Hamilton's Bar**.

Fahy Bay

Latitude: 53°33'.50N; Longitude: 10°00'.60W;
Charts: Admiralty 2420 2706; Imray C54

Leaving Clifden passage can be made through High Island Sound. The inhabited island of Inisbofin lies to the north. Ballynakill Harbour provides excellent shelter and is entered rounding Cleggan point. Note the Carrickmahoy rock between it and Inisbofin. There are eight visitor mooring buoys in the delightful anchorage of Fahy Bay.

Inisturk

Latitude: 53°42'.30N; Longitude: 10°05'.20W;
Charts: Admiralty 2667 2706; Imray C54

Inisturk lies about eight miles from Roonah Point and ten miles from Cleggan, and is served by ferries from both locations. It has a population of just under 100 and is renowned for its scenery, cliff flora and bird sanctuaries. It has a rugged landscape ideal for walking tours. Eight visitor mooring buoys have been placed near the harbour. There are several guest houses which provide meals and meals are also served in Delia Concannon's by the harbour.

Clare Island

Latitude: 53°48'.10N; Longitude: 09°56'.80W;
Charts: Admiralty 2420 2667; Imray C54

Clare Island, about four miles in length, lies about two miles off Roonah Point, at the entrance to Clew Bay. It has the best farming land of any island off Ireland's west coast, with fields defined by stone walls. Knockmore Mountain (500 metres) has a precipitous descent down to the water's edge. It is a peaceful haven for walking and reflecting, far from the frenetic pace of western daily life. On the east of the island is the harbour where there are three visitor mooring buoys and from which there are regular ferry sailings to Roonah Quay.

Clare Island's best known inhabitant was Gráinne Ni Máille, Grace O'Malley alias Granuaile. The exploits of this lady, a fearless sea captain, both on land and sea, with romantic embroidering, have become the stuff of legend (see entry on Howth). She fought off an English attack on her mainland stronghold and still had time to acquire two husbands. Her second husband Richard Burke, organised a rebellion against the English. Gráinne was arrested and taken to England. Such was her reputation as a fearsome leader, that she was able to meet and petition Queen Elizabeth I, who pardoned her, on her accepting a licence to harry the Queen's enemies. She is reputedly buried on Clare Island. The remains of her castle overlook the harbour.

About a half mile east of the harbour is St. Brigid's Bed, with a covered holy well. A mile and a half to the south west are the remains of a Cistercian Abbey which has rare remains of frescoes. The island also has the remains of two promontory forts.

Looking to the west Croagh Patrick dominates the skyline, 800 metres in height. This is Ireland's holy mountain, where tradition tells us that St. Patrick fasted and prayed, ringing a bell which banished snakes from Ireland (the local wits say that they swam the Atlantic and joined the Mafia!). Every year on the last Sunday of July, thousands, many in bare feet whatever the weather, make the penitential climb to the summit. This is known as the 'Croagh Patrick Pilgrimage'. Mass is celebrated in a tiny church which can be seen from some miles away, perched on the mountain top. First aid teams have a busy day, with the inevitable casualties from falls, many of them serious with many other pilgrims suffering from sheer exhaustion.

At the base of Croagh Patrick is the village of Louisburgh. During the Great Famine, 1845 to 1849, the province of Connaught and particularly Mayo suffered enormously. A diet consisting mainly of potatoes, normally kept people healthy and disease free. When blight struck the crop, the population was left in a disastrous situation. As a result of the death and emigration, Ireland's population was decimated. The national monument to the famine is located near Louisburgh. It consists of a striking bronze replica of the type of ship which took famine victims to a new life across the Atlantic.

Clare Island has one hotel, The Bayview, which has a bar and restaurant. The Community Centre also has a bar, and can provide showers. There is a coffee shop at the harbour. There is one shop about two miles away. Bicycles can be hired at the harbour.

Blacksod

Latitude: 54°06'.30N; Longitude: 10°03'.70W;
Charts: Admiralty 2420 2704; Imray C54

To the north west of Clare Island lies Achill, Ireland's largest island. It just manages to be an island, joined as it is now to the mainland by a short bridge. It is a very popular holiday location with fine mountain and cliff views. Passing from Clare Island to Achill, to the west lies Clew Bay with an enormous scattering of islands. Around Achill Head to the north is the Belmullet Peninsula.

At the southern tip of the low-lying peninsula lies Blacksod Quay with its lighthouse. The Commissioners of Irish Lights maintain a helicopter landing pad here from where servicing flights are made to other lighthouses and installations. Six visitor mooring buoys are located to the south of the quay. A landing can be made either at the quay or at a pontoon a short distance to the west. The harbour is used as a base for anglers setting out on day trips. There are no facilities here apart from a phone kiosk. Bar meals are served in An Currach Bar about four miles away in Elly Bay.

Elly Bay

Latitude: 54°10'N; Longitude: 10°03'.60W;
Charts: Admiralty 2420 2704; Imray C54

About four miles north of Fallmore Point on the eastern side of the Mullet Peninsula is Elly Bay where there are six visitor mooring buoys. Unfortunately the buoys are located about a half mile from the shore. There are no facilities apart from An Currach Bar which serves bar food.

Ballyglass

Latitude: 54°15'.20 N; Longitude: 09°53'.50W;
Charts: Admiralty 2420 2703; Imray C54

Up to some years ago it was possible to travel by the Belmullet Canal from Blacksod Bay into Broad Haven. It is now closed. Ballyglass Pier is a half mile west of Broad Haven Lighthouse. Rounding Erris Head to enter Broad Haven the uniquely shaped Broad Haven Stags are visible. These rocks viewed from a distance could be enormous spinnakers! The Ballyglass Lifeboat is stationed here. The pier is usually busy with trawlers and angling boats coming and going. There are eight visitor mooring buoys unfortunately located about a half mile to the south west of the pier. There is a fresh water tap on the pier and a public phone behind the Lifeboat Station. The nearest facilities are in Belmullet, about six miles away with shops, Post Office, bank, restaurants and pubs.

Kilcummin

Latitude: 54°17'.30N; Longitude: 09°14'.80W;
Charts: Admiralty 2715 2767; Imray C54

From Ballyglass around Benwee Head is the north Mayo coast. Just inland of Downpatrick Head a remarkable archaeological discovery known as the Ceide Fields has recently been made. This area now is bogland, but 5000 years ago migrant Celtic farmers established themselves here and cleared the forests covering the land. Forests will retain 70 per cent of the rain falling on them, only 30 per cent reaches the ground. A combination of this tree clearance and perhaps a changing wetter climate eventually turned this pasture land into bog. In certain wet conditions, plants do not decay but turn into peat moss. More plants grow on top, and the process repeats itself, with matter below being compressed. Over the millennia a dense bog develops. The peat was and is still harvested, dried and used as fuel. In digging, farmers in recent year questioned as to how there were stones a couple of metres beneath the surface. They seemed to form a pattern, which indeed they did. Archaeologists came and started a detailed survey. They discovered ten square kilometres of walled fields. The walls had been laid out along the land contours to facilitate the enclosing of level fields. Remains of houses, cattle pens and burial plots have been uncovered. The Celts were a well organised people, certainly of a peaceful nature, as no weapons have been uncovered. This is the largest Stone Age relic extant in Europe. A gracefully designed Interpretative Centre houses an exhibition, describing the origins of the Fields and their inhabitants. The Centre has won architectural praise, its lines clearly visible on the skyline and imaging the Stags of Broad Haven out to sea.

Kilcummin Head is at the eastern entrance to Killala Harbour, a picturesque stretch of water leading to the town of Killala. Nearby is Kilcummin Quay close to which eight visitor mooring buoys have been laid. The town of Killala is about six miles away by road.

This quiet hamlet was the scene of a major episode in Irish history. Above the pier is a monument with the inscription 'Voici l'endroit ou Le General Humbert a debarqué avec des soldats Francais le 22 Aout 1798'. The French in response to a request from Theobald Wolfe Tone supplied 700 troops, to assist in the 1798 rebellion. Wolfe Tone, a Presbyterian, inspired by the American and French revolutions, visited both countries, and was the driving force behind this rebellion against English rule. Given what subsequently transpired in Northern Ireland, it is of great interest that the roots of modern day Irish republicanism emanate from Tone, who wanted a state in which the common name of Irishman replaced the denominations of 'Protestant, Catholic and Dissenter', and all could work and live in peace together without harassment.

Humbert marched to nearby Castlebar, and defeated a larger force of English troops. The English retreat was subsequently dubbed 'The Castlebar Races'. Humbert pressed on through Sligo and Leitrim, until he was confronted by an English army, commanded by Lord Cornwallis at Ballinamuck, Co. Longford. Here Humbert was defeated and the survivors put to the sword. Tone himself attempted to land with a larger force at Lough Swilly, but was overpowered. Taken prisoner, he was transported to Dublin, tried and sentenced to death. Rather then face the gallows he took his own life, thus ending another sad episode in the Irish search for self-determination.

A plaque at the quay commemorates the landing and was placed by a famous Irish nationalist, Maud Gonne McBride. She was a renowned beauty, 'with the walk of a queen' whose hand the poet William Butler Yeats unsuccessfully sought. Her husband, Major John McBride, who had fought with the Boers in South Africa against the British, was subsequently executed for his part in the 1916 rebellion. Her son, Sean McBride was a noted politician, who received the Nobel Peace Prize for his work in assisting the United Nations in the establishment of Namibia as an independent state. Interestingly, in contrast, he presided over the enforced disengagement of South Africa from the affairs of Namibia.

There are few facilities in the area, apart from **Bessie's Bar** a short walk away. About a mile towards Killala is another bar, **The Kerryman's**. Killala about six miles distance by road has the usual range of shops, a Post Office and bank, friendly pubs and eateries.

Teelin

Latitude: 54°37'.50N; Longitude: 08°37'.80W;
Charts: Admiralty 2702 2792; Imray C53

Across the beautiful stretch of Donegal Bay, to the north east lies Teelin Harbour, in a small inlet about half way between Rathlin O'Beirne Island, and the fishing port of Killybegs. Nearby is Slieve League whose cliffs rise an awesome 600 metres from the sea, making them the highest marine cliffs in Europe. Teelin in the days of sail, flourished as one of Ireland's largest fishing ports. It was built in the 1880s through the generosity of Lord Bradford who was a regular visitor to the area. With the advent of steam power, boats moved further up to Killybegs, which today is Ireland's largest fishing port with major fish processing facilities. There are the remains of what was a large coastguard station, built about 1813, but destroyed in 1921 during the War of Independence.

Four visitor mooring buoys are located about a cable from the end of the pier, which has fresh water. There is enough water to tie up towards the outer end of the pier. The town of Carrick is the nearest supply centre, about three miles away, with a Post Office, McGinleys general store, and a taxi service, Carrick Cabs. There is a fish processing factory on Teelin Pier and the owner Jack Gallagher and his friendly staff will help with queries regarding diesel and other supplies.

Pubs and restaurants

The Rusty Mackerel Pub, a lively spot in the evening is less than a mile away.

The Slieve League Bar and **Enright's Bar** in Carrick both serve pub food.

Portnoo/Iniskeel

Latitude: 54°50'.82N; Longitude: 08°26'.50W;
Charts: Admiralty 1879; Imray C53

From Teelin the journey to Portnoo is around Rossan Point (a reference location in Met Eireann weather bulletins) and Dunmore Head. Portnoo lies on Gweebarra Bay noted for spectacularly beautiful beaches and scenery. Six visitor mooring buoys have been placed off the island of Iniskeel, with its ruined churches, and early Celtic engraved stone slabs. Iniskeel can be reached by foot at low tide as the sandbank dries. The village of Narin is about a half mile away, but the dinghy would need to be hauled to dry land, across the strand, away from the water. There is a pier about three quarters of a mile to the east at Portnoo which has a Post Office and Morgan's grocery store. In Narin there is the **Narin Bar** which serves food and **Andy's Cafe and Takeaway**. There are public phones on the beach.

Aranmore

Latitude: 54°59'.50N; Longitude: 08°29'.40W;
Charts: Admiralty 1879 1883; Imray C53

Aranmore Island lies off the coast of Donegal about two miles from the fishing port of Burtonport. This part of Donegal is known as The Rosses. With many lakes and streams it is an angler's paradise, but is relatively poor farming land. The Irish language is widely spoken. The island about three miles by three miles is well populated. A ferry runs regularly to Burtonport threading its way through channels between the islands which guard the entrance. The numerous islands with their narrow channels have led this area to be dubbed the 'The Venice of the North West'. An attempt was made to build a fishing port on nearby Rutland Island but the wind and sand reclaimed it and nothing much is left. Aranmore has spectacular cliffs and caves on the western side.

Six visitor mooring buoys are located to the east of the village on the eastern side of the island close to the lifeboat mooring. There is a slip about a quarter mile row from the moorings. The village is about a half mile walk. There is a grocery – Boyles and a Post Office. There is a good restaurant attached to the Ferry Guest House – Bonners. There is also a small hotel, The Glen.

Downings

Latitude: 55°11'.30N; Longitude: 07°50'.30W;
Charts: Admiralty 2699; Imray C53

Around Bloody Foreland (another Met Eireann weather forecast reference location), and Horn Head lies Sheep Haven. This is rugged and isolated but scenic coastline. To the north lies Tory Island, the most remote of Ireland's offshore islands. Tory can be isolated for weeks on end because of Atlantic storms, although helicopters are now able to reach it for emergencies. Various moves made to relocate islanders to the mainland have been resisted, so it still retains a sizeable population, which in recent years is in fact increasing. There is a regular ferry service from Downings.

The latter is located on the eastern side of Sheep Haven Bay. Nearby is Rosapenna with a famous championship golf course, a links. This is a very popular holiday destination with great rolling golden sand dunes. There are wonderful sandy beaches and sea and mountain vistas. Eight visitor mooring buoys have been laid about two cables from the end of the pier. The pier has toilets and fresh water. The local fishermen are helpful and will advise about obtaining diesel and other supplies. There is good water at the outer end of the slip. The Post Office and public telephone are located a short distance from the pier.

About a half mile away can be found shops and pubs:

Bradley's and McGiolla Brighd grocery shops.

Downings Bar, **Tramore Inn** which serves food, **The Beach Hotel and Restaurant**. There is also a takeaway.

Portsalon

Latitude: 55°12'.50N; Longitude: 07°37'W;
Charts: Admiralty 2697; Imray C53

Leaving Sheep Haven and rounding Fanad Head brings you to Lough Swilly. It is entered between Fanad and Dunaff Heads. Superlatives have also been used to describe the scenery here. Lough Swilly, over twenty miles in length, has cliffs, wooded shores and magnificent sweeping sandy beaches. About four miles from the entrance on the western side is Portsalon. The name probably derives from the fact that salt was at one time imported here – in Irish, Port na tSalainn, the Salt Harbour.

Lough Swilly has many historical connections. The Irish republican leader, Wolfe Tone, landed here in 1798 with a party of French soldiers to aid the rebellion. They were overpowered and Wolfe Tone arrested, and moved to Dublin to be tried for treason (see entry on Kilcummin). It is a place of shipwrecks too. The frigate *Saldanha* was wrecked in 1811, and the nearby headland is called after it. In 1917 the bullion laden White Star Liner, *Laurentic*, was sunk by a German U-boat off the mouth of the Lough. The bell of the *Laurentic* was recovered by divers and is now in the local Protestant church. Over the years much of the bullion has also been raised.

Lough Swilly was one of the three ports which remained in British hands, under the 1921 Treaty of Independence. Britain ceded the ports in 1938. There are remains of several fortifications in the Lough built to resist Napoleon, who as it turned out, fortunately focused his attentions elsewhere. One of them, Fort Dunree on Dunree Point, was an Irish army post until recently and now houses a museum.

A half mile walk will bring one to Ballydaheen House and Gardens. The owners Mary and David Hurley lived for a time in Japan, so they chose to build the house and lay out the gardens in Japanese style. Visitors are welcome to explore the gardens with their superb selection of trees, shrubs and flowers imported from many parts of the world. Newly married brides come to be photographed on the Japanese bridge. A donation to the Lifeboats gains entrance. On a fine day the views are stunning with Dunaff Head in the distance.

There are eight visitor mooring buoys laid close to the pier at Portsalon. Nearby is a sheltered beach where children can safely busy themselves with the sand and sea. There is also a fine and larger sweep of sandy beach to the south.

There is a grocery and bar at the pier untill recently presided over by the legendary Rita Smyth, Queen of Fanad, who has sadly passed away. Next door is a simple restaurant. Meals are also served in the golf club which is a short walk away. The village at the cross-roads is a half mile walk, where there are two general stores. Here also is the Ardglass Inn.

Rathmullan

Latitude: 55°05'.70N; Longitude: 07°31'.70W;
Charts: Admiralty 2697; Imray C53

A further four or so miles up Lough Swilly also on the western side, lies the town of Rathmullan. There is a quay here where factory trawlers and other commercial vessels can unload. The Rathmullan Enterprise Group have installed a pontoon, with power and water, on the southern side of the quay. It can accommodate up to 20 boats. The charge is £5 for a short stopover and £10 for an overnight stay. This is a fine example of local initiative which has brought plenty of businesss to the town. Skippers should report to the friendly Pier Hotel, who if not busy, can organise showers. They will answer any queries and advise on diesel supplies and repairs. They can put boat owners in contact with a local marine engineer, James Montgomery.

On the shore is Rathmullan Fort, now restored with swivelling gun emplacements uncovered after years of neglect. The fort houses a Heritage Centre and the visitor will be rewarded with a fascinating insight into the history of the island of Ireland. For it was from this point that, in 1607, the Ulster chieftains O'Neill and O'Donnell left Ireland, hoping to return one day to regain their possessions. But it was not to be. This episode is known in Irish history as 'The Flight of the Earls' (see entry on Kinsale). It signalled the eclipse of Gaelic Ireland and completed the English conquest of the island.

With the Viking and subsequent Anglo-Norman conquests, Ireland was effectively divided. The Norman descendants, Burkes, Fitzgeralds and Ormondes largely controlled the east and south, under the English crown. Gaelic Ireland with its own legal system of Brehon Laws and traditions that gave poetry and language a special position flourished in Ulster. During the reign of Queen Elizabeth I, most of the country had been subjugated, its leaders defeated, English law and the Protestant religion imposed. Only Hugh O'Neill and Red Hugh O'Donnell held out.

In 1601 Spain under Phillip III sent a fleet to Kinsale (see entry on Kinsale) to assist O'Neill and O'Donnell. The mission was a failure and O'Donnell left Ireland. He was succeeded by his brother Rory O'Donnell. But eventually they too gave up the struggle and left.

Any remaining smaller chieftains were defeated and the plantation of Ulster began; the land was in effect confiscated and reallocated. The new residents, known as undertakers, were obliged under oath to recognise the sovereign power of the king and to accept the Protestant religion. Celtic feudalism and Brehon laws were replaced by English and Scottish landlords and laws. The subsequent development of the political, social and economic life of Ulster is described, including the detailed map-making required to delineate the new order as a market economy replaced the older trading methods. New towns were planned. There is a scale model of Derry, the second city of Northern Ireland (or Londonderry, there is still a division today over its name).

James I granted the lands and city of Derry to the citizens of London who planned a new city to be surrounded by walls and planted it with settlers. The city fiercely defended a blockade by the Jacobites in 1689. The siege was lifted on August 12th, and on that day every year, the closing of the gates against the oppressors by the Apprentice Boys is commemorated in a parade. The city was renamed Londonderry to commemorate the link. Visitors wishing to understand the origins of the troubles which have beset Northern Ireland in recent years, should find some clues here. But as one of the exhibits so rightly states, it is better to read the pages of history and to try to understand them, rather than tear them out.

Restaurants and pubs:

The Pier Hotel has a restaurant and also serves bar food, tel: 074 58178/58115.

The Ferry Gate restaurant is open Thursday to Sunday.

Rathmullan House Hotel nearby has a good restaurant and also a leisure centre including an indoor swimming pool.

You will also find a Post Office and two grocery shops.

Culdaff

Latitude: 55°18'N; Longitude: 07°09'.10W;
Charts: Admiralty 2697; Imray C53

The great Inishowen Peninsula reaches out into the Atlantic with Lough Swilly to the west and Lough Foyle to the east. Malin Head at its tip is the most northerly point on the island of Ireland. Further north still is the uninhabited Inishtrahull Island. Culdaff lies on an inlet on the north east side of the peninsula. There are six visitor mooring buoys about one cable east of the pier. The village is located about a mile away.

McGrory's Restaurant, tel: 077 79104.

Moville

Latitude: 55°11'N; Longitude: 07° 03'W;
Charts: Admiralty 2499; Imray C53

Leaving Culdaff, Inishowen Head lies about seven miles to the south east, and once around it, Lough Foyle opens up, leading all the way to Northern Ireland's second city, Derry. The entrance to the Lough proper is between the busy fishing port of Greencastle to the north west and Magilligan Point to the south east. Clearly visible especially at night when floodlit is Magilligan Prison. About three miles past Greencastle lies Moville, which has eight visitor mooring buoys, laid about three cables south west of the pier head. There are no facilities in the area.

Coleraine

Latitude: 55°10'.30N; Longitude: 06°46'.35W
Charts: Admiralty 2499; Imray C53

Coleraine in Co. Derry lies about five miles from the mouth of the River Bann. Care should be taken entering and travelling up the river. Some reports say that since dredging stopped, serious silting has taken place.

Coleraine was one of the towns planned following the plantation of Ulster. To the rear of St. Patrick's Church in the town can be seen the remains of an earthen wall or rampart, which surrounded the town in the early 1600s. Coleraine developed as a centre for the processing of linen, made from locally grown flax. This was once a major industry, but has declined although Northern Ireland linen is still very much in demand as a unique material for table cloths, napkins and cool dress fabrics, possessing as it does a special sheen or lustre. Today it is a busy commercial centre with the University of Ulster nearby. There is a 55-berth marina, near the town which lies between the pleasant grassy banks of the River Bann. Charges for an overnight stay range from £7 to £9 depending on length. The marina is fully equipped with power, water, toilets and showers and diesel and petrol are available. There is a fourteen-ton travelling hoist. Marine repairs are available.

Next door is Coleraine Yacht Club which welcomes visitors to its bar in the evening. The town of Coleraine can be reached by a bus, which passes the entrance of the yacht club approximately every half hour. Alternatively, it is a quarter mile row up the river where you can tie up close to Dunnes and Tesco supermarkets. There is a Post Office and a range of banks and shops.

Several restaurants to chose from include:
Water Margin, 42 High Street, tel: 42222 **Little Caesar**, 45 Railway Road, tel: 329991 **Sun Do** (Chinese Takeaway).

An alternative is to take the bus to the nearby popular coastal resort of Portstewart. Plans are well advanced for the construction of a marina here. Another few miles brings one to Portrush, probably Northern Ireland's most popular holiday destination. It has a delightful seafront for strolling. The coast here is punctuated with some of Ireland's finest golf courses, including Royal Portrush, Portstewart (two courses), Strand, Castlerock and Ballycastle.

Portstewart also has a promenade, again for pleasant seaside strolls along which there is a good selection of bars and restaurants:

The Montague, 68 The Promenade, tel: 834146, **Ashianna Tandoori**, 12a The Diamond, tel: 834455, **The York Bar**, **The Anchor Bar**.

Ballycastle

Latitude: 55°12'.50N; Longitude: 06°14'.30W;
Charts: Admiralty 2798; Imray C53

Further east along the beautiful Derry and Antrim coasts lies Ballycastle. Nearby is the world famous 'Giant's Causeway' with its geometric patterns of volcanic basaltic rock. There are tens of thousands of hexagonal stone columns here, not manmade but the result of cooling molten lava. On a promontory at Portballintrae are the spectacular ruins of Dunluce Castle, originally built by Richard de Burgo. It was fought over by the MacQuillans and MacDonnells eventually coming under the Earls of Antrim. Not far away is the village of Bushmills, a household name as a result of the whiskey (or to use the Gaelic – Uisce Beatha – the Water of Life), distilled here. It was founded in 1608 and is reputedly the oldest whiskey distillery in the world. Rathlin Island is also clearly visible. Normally a quiet haven, it was projected onto the world's tv screens when Richard Branson of Virgin fame ended his attempt to circumnavigate the earth by balloon near the island. The islanders played a major part in the rescue. Branson generously expressed his thanks by presenting the island with a high-speed launch and a community centre at Ballycastle.

Moyle County Council has recently completed a fully equipped 70-berth marina at Ballycastle, complete with power and water (tel: 012657 62024). Daily rates are £10 up to 9 metres rising to £16 for 12 metres and over. Visitors should report to the marina office, which has a laundry attached. Ballycastle is the terminus for ferries to Rathlin Island and also to Campbeltown on the Mull of Kintyre in Scotland.

Part of Ballycastle lies on the seafront with the older part about a mile inland. This is the location for The Auld Lammas Fair, which takes place in August, with centuries of tradition. About a mile to the east at Bunamargy, lie the ruins of a Franciscan Friary, which was sacked in 1584. It was subsequently used as a burial ground for the MacDonnells and their heirs the Earls of Antrim. In the distance is Fair Head with its distinctive shape sloping down into the sea.

There is a Post Office and grocery shop on the seafront. There are also several pubs and restaurants:

Restaurants: (all in North Street): **The Strand**, tel: 028 2076 2349; **No 10**, also has a wine bar, tel: 012657 68110; **The Marine Hotel**.

Bars: Marconi's Bar, **The Castle Bar**, **The Angler's Arms** and the **Harbour Bar**.

A ten minute walk brings you to the town of Ballycastle proper. For restaurants there is **Wyseners's**, an award winner, tel: 028 2076 2372; also **Connolly's/Anzac Restaurant**, Bar and Wine Bar and Pub serving bar food.

Refer to map on page 99.

Carnlough

Latitude: 54°59'.87N; Longitude: 05°59'.20W;
Charts: Admiralty 2198; Imray C64

Rounding Fair Head the Antrim coast is dotted with towns like Cushendun and Cushendall. Inland are the celebrated and beautiful Glens of Antrim. From Cushendall and all the way to Belfast, the Antrim coast road provides drivers with one the most panoramic drives in Ireland. The coast of Scotland is clearly visible across the North Channel.

Carnlough is a small fishing port on this coast, which is a useful stopping off point for sailors heading to or coming from Scotland. It is used by fishing boats; visitors should be able to tie-up alongside. Local stone was shipped from here, and the bridge which carried the railway into the harbour can be seen crossing the main street. Fresh water is available. There is a bank, Post Office and supermarket. There are several restaurants and pubs along the seafront, including a Chinese takeaway:

Londonderry Arms, an ivy clad hotel with restaurant and bar, tel: 01564 885255; **Harbour Lights** serving lunches and teas, **The Glencloy Inn** and **McAuley's Bar**.

Carrickfergus

Latitude: 54°42'.70N; Longitude: 05°48'.20W;
Charts: Admiralty 1753; Imray C62

Leaving Carnlough, the Maidens, a group of rocks are to the south east. The mouth of Larne Lough is to the south. This is a busy ferry port with conventional and high speed ferries crossing to Stranraer and other ports in Scotland. Around Island Magee and Black Head with its lighthouse lies Carrickfergus on the northern shore of picturesque Belfast Lough.

One building which dominates the waterfront is the magnificent Carrickfergus Castle, built by either John de Courcy or Hugh de Lacy. This Norman castle was captured for the crown by King John in 1210. It remained a loyal English fortress with minor interruptions. It was here that King William of Orange first landed in Ireland in 1690 on his way to victory over the Jacobites at the Battle of the Boyne. This is commemorated both by a statue and by an annual re-enactment of the landing. The date was the twelfth of July and every year on that day thousands of Orangemen, in numerous towns of Northern Ireland, march to the sound of the Lambeg drums to commemorate the battle.

The castle is a national monument open to the public and has undergone extensive restoration. It was briefly captured by the French in 1760, and in 1778, the first action by a warship of the newly founded United States Navy, took place off the coast, when the *Ranger* commanded by Paul Jones defeated the British Man O'War *Drake*.

There are also other reminders of the United States: the parents of Andrew Jackson, the seventh president, hailed from Carrickfergus. There is the Andrew Jackson Heritage Centre housed in a thatched cottage of the period. And during World War II US troops stationed in the area trained for their role in the D-Day invasion of France. Other prominent names associated with Carrickfergus are Jonathan Swift who spent some time here ministering in a local parish, and this was also the birthplace of Louis McNeice the poet. There has been a great boatbuilding tradition here, and it was in Carrickfergus that the Howth 17 footers were built (see entry on Howth for more details).

Carrickfergus has a 300-berth marina, the entrance has leading lights. The pontoons have water and power, and holding tanks can be emptied. There is a 45-ton travelling hoist. Full shore facilities are available; toilets, showers and a launderette. All boat, engine and sail repairs can be undertaken. Daily berth charges are £3 per metre (£1.50 for overnight stay) and £7.50 per week.

Within the marina complex which is in the final stages of development there is a supermarket and Post Office. A five minute walk brings one to High Street with banks and shops.

Restaurants and bars:

The Wind-Rose restaurant in the marina area has a bar and serves a wide range of meals, tel: 01960 364192.

Dobbins Inn Hotel, 6 High Street, with the **De Courcy Restaurant** and **Paul Jones Bar**, tel: 01960 351905.

Chandler's Restaurant and Wine Bar, 13 High Street, tel: 01960 369729.

The Quality Hotel, 75 Belfast Road, **Macs Bar**, **Boardwalk Restaurant**, tel: 01960 364556.

Refer to map on page 100.

Bangor

Latitude: 54°40'N; Longitude: 05°40'W;
Charts: Admiralty 1753 2198; Imray C62

The town of Bangor lies about 12 miles north east of Belfast, beside Ballyholme Bay on the southern shore of Belfast Lough. It has long been a popular resort with Belfast residents. Its name derives from the Gaelic for 'curved horns' – Beannchor, describing the shape of the bay. Bangor marina has 560 berths. Visitors should call the marina office on Channel 80 before entering. If the red 'traffic lights' are lit (three vertical) contact the duty berthing master. The entrance between the Northern and Pickie breakwaters is clearly marked. Visitors should report to the office on arrival and obtain the PIN number which will open the entrance gate. The marina has all services; 240v power, water and fuel. There is a shower block with laundry attached. Daily rates are £1.50 per metre per day (minimum charge £10), £8.75 per metre per week.

Apart from some Norse artefacts in The National Museum in Dublin, very little remains of early Bangor. It flourished as one of the great centres of European Christian learning from the 6th century. And it was from here that Saints Columbanus and Gall, having studied under St. Comgall, sailed to relay the Christian message to Europe. But inevitably it fell prey to Viking invaders, its monasteries sacked and the inhabitants killed or scattered.

In the Victorian era, Bangor grew up as a prosperous town close to Belfast with thriving shipbuilding and engineering industries. Its history can be studied in the heritage centre in the buildings attached to Bangor Castle. The Ulster Folk and

Transport Museum with displays of Ulster life as well as railway and road transport is just a few miles from the town.

In 1912 with the passing of the Home Rule Bill, Ulster Unionists under their leader Sir Edward Carson, became increasingly concerned about being absorbed into an independent Ireland. Home rule would be 'Rome rule', was the war cry, reflecting their views as to who the real ruler of Ireland would be. An Ulster Volunteer Force was recruited and in 1914 they landed cargoes of guns at Larne, Donaghadee and Bangor. But unlike future events in Dublin, there was to be no general uprising in Ulster.

Ireland's second largest city – Belfast, on the river Lagan, is served by buses and trains from Bangor. Its name derives from the Irish – béal feirste, the mouth of the sand bank. An Anglo-Norman – John de Courcy built his castle in 1177 near what is now Castle Place to command the river. The Irish and Anglo-Normans frequently clashed, but their lands were eventually confiscated and bestowed on the Chichester family. Lowland Scots arrived as part of the plantation of Ulster. In the late 17th century there was an influx of Huguenots fleeing France, after the Revocation of the Edict of Nantes. They established a thriving weaving industry which led to the development of the linen trade. Whereas Dublin is essentially Georgian, Belfast is in contrast a Victorian city. For it was in the 19th century that it became a major industrial centre. Shipbuilding, textile machinery, tobacco and of course linen manufacturing thrived. The depression of the 1930s and the rise, after World War II, of Asian competition led to a decline. The civil rights upheavals from 1968 and bitter sectarian strife all took their toll.

But today with the peace process taking firm root, Belfast is once again a thriving metropolis. The city centres itself around the imposing City Hall on Donegall Square with the main thoroughfares Donegall Place and Royal Avenue leading off it. Other fine buildings include St. Anne's Cathedral, Grand Opera House, Albert Memorial Clock Tower and the Linenhall Library. Another landmark is the enormous crane of the Harland and Wolff shipyard, visible for miles. A magnificent new concert hall has recently been opened on the waterfront, one of many new buildings. There are frequent flights from Belfast International and City Airports to destinations in mainland Britain, Dublin and Shannon. Rail and bus services connect with Dublin, Derry, Coleraine and Newry.

A must in Belfast is a visit to the Crown Liquor Saloon, 44 Great Victoria Street, featuring wonderful stained glass, marble and carved oak. There are many other pubs and eateries – the best known restaurant is Roscoff's at Shaftesbury Square, tel: *331532; others include Nick's Warehouse, 35 Hill Street, tel: *439690; Antica Roma, 67 Botanic Avenue, tel: *311121 (*Belfast code is 01232).

Bangor has long been a great yachting centre. A little over a mile from the marina is the imposing building of The Royal Ulster Yacht Club, which welcomes visiting yachtsmen to its bar and dining room. And close by right on the seafront is Ballyholme Yacht Club, again welcoming visitors. Both clubs have extensive series of racing events during the season.

Bangor is a busy shopping town with banks, Post Office and usual facilities. There is a wide range of hotels, restaurants and bars:

Marine Court Hotel, opposite the marina with the **Stevedore Restaurant**, **Lord Nelsons Bistro** and **Calico Jack's Bar**, tel: 01247 451100.

Bangor Steak House, 119 High Street. tel: 01247 470768.

Jenny Watts Bar, 41 High Street, Victorian interior, bar food 'til 2130, jazz sessions Sunday lunchtime, tel: 01247 270401.

Donegan's Bar and Restaurant, 44 High Street, tel: 04981 270362.

Knuttel's Restaurant, 7 Gray's Hill, does not have a licence, but customers can bring wine and there is no corkage charge, tel: 01247 274955.

McMillen's Bar and Restaurant, 30 Quay Street, tel: 01247 467699.

Shanks, this is a Michelin one star restaurant but expensive, 150 Crawfordsburn Road, Helen's Bay (a three mile taxi ride), tel: 028 9185 3313.

Genoa Bistro, 1a Seacliff Road, tel: 01247 469253.

Refer to map on page 100.

Kerry - between Ventry and Smerwick

Maps

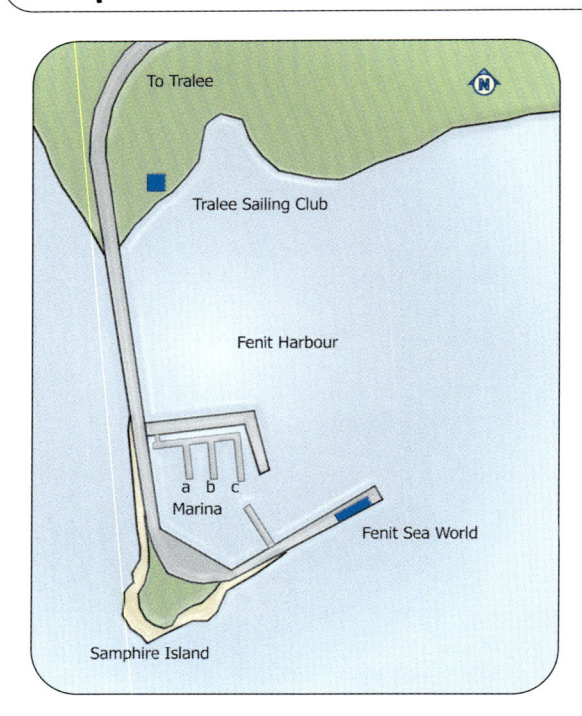

Fenit
(see page 69)

Map showing: To Tralee, Tralee Sailing Club, Fenit Harbour, Marina (a b c), Fenit Sea World, Samphire Island

Portsalon *(see page 87)*

Maps

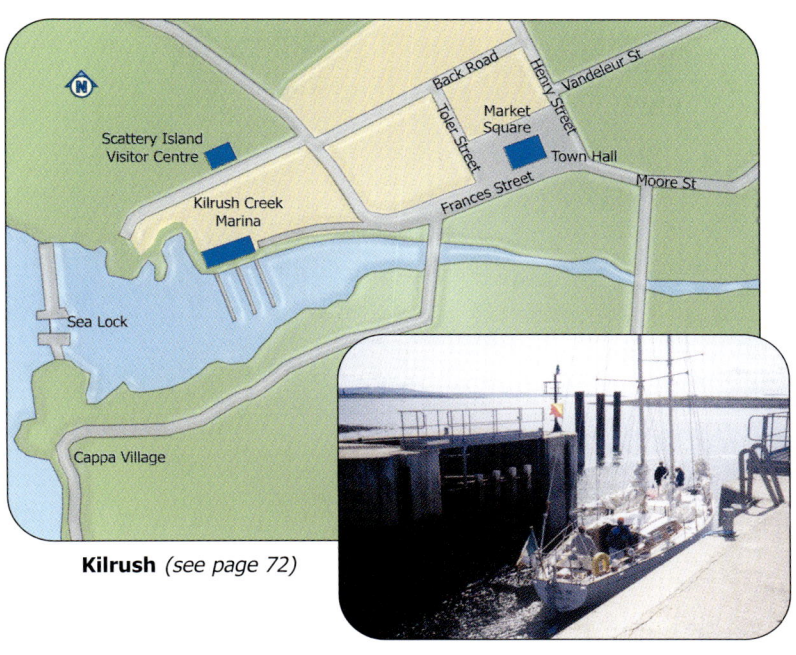

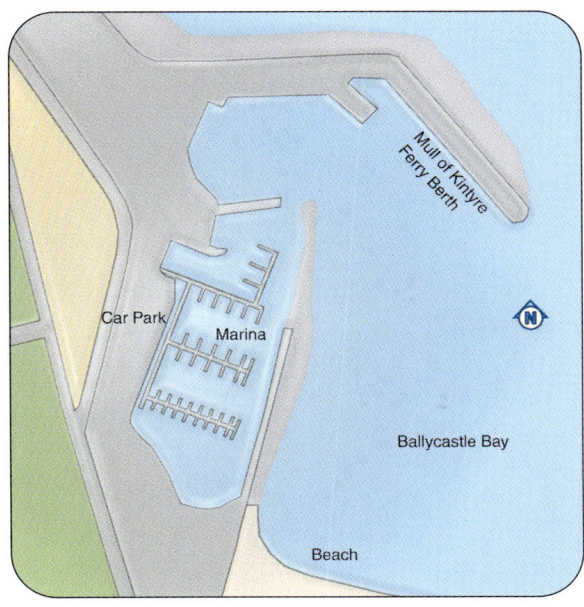

Kilrush *(see page 72)*

Ballycastle
(see page 92)

99

Maps

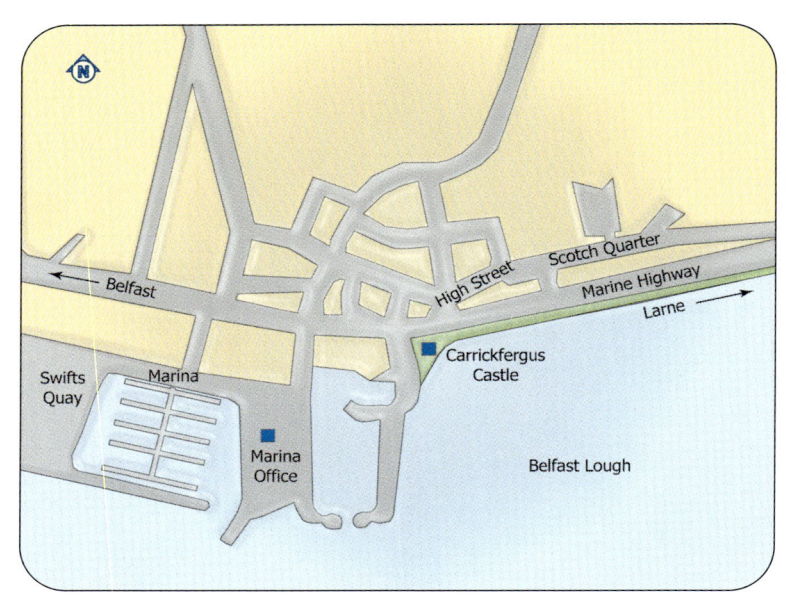

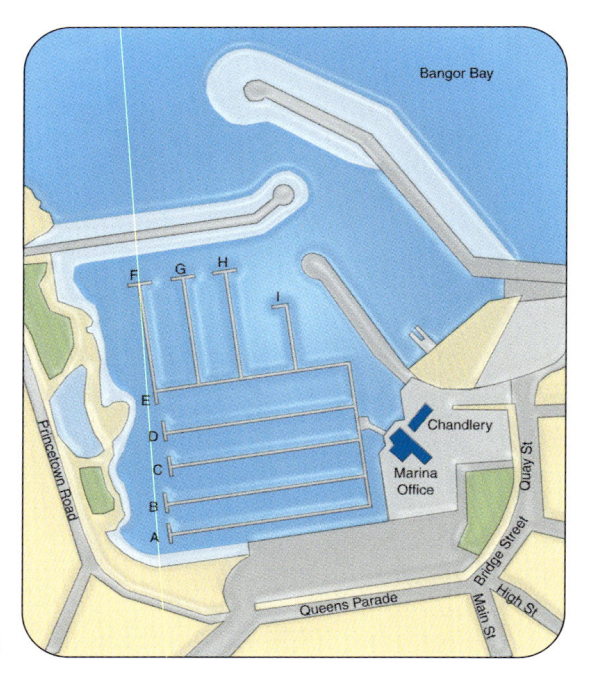

Carrickfergus
(see page 94)

Bangor
(see page 95)

Maps

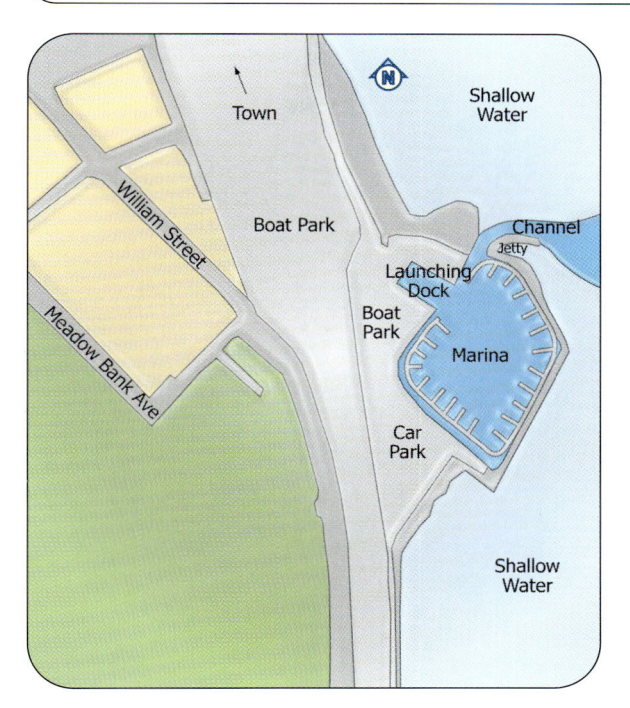

Donaghadee
(see page 105)

Ardglass *(see page 107)*

Maps

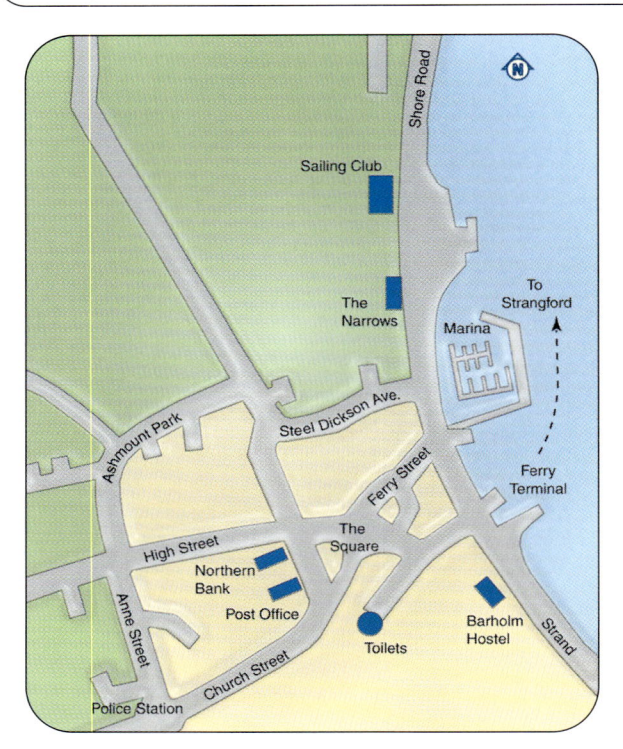

Strangford Lough *(see page 106)*

Portaferry Marina *(see page 106)*

Maps

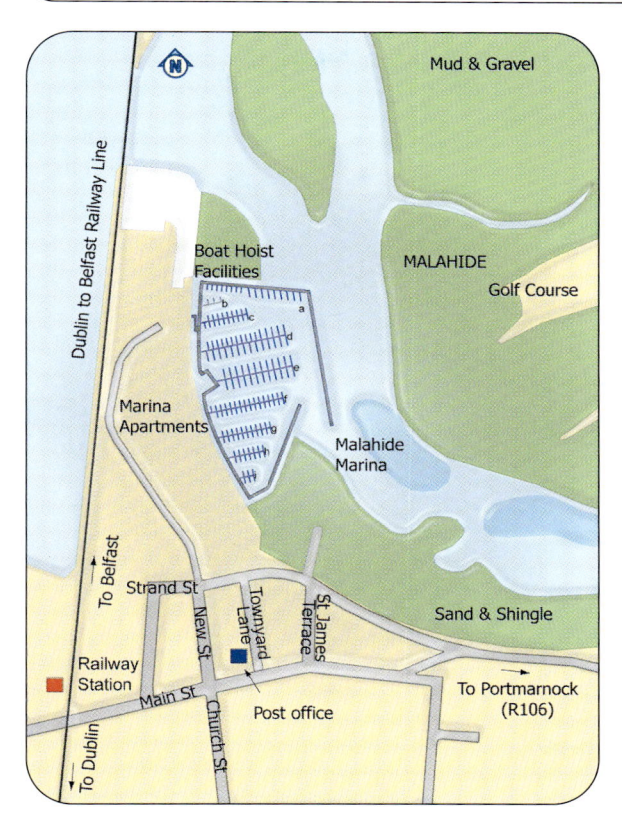

Malahide
(see page 109)

Clifden Bay

103

Maps

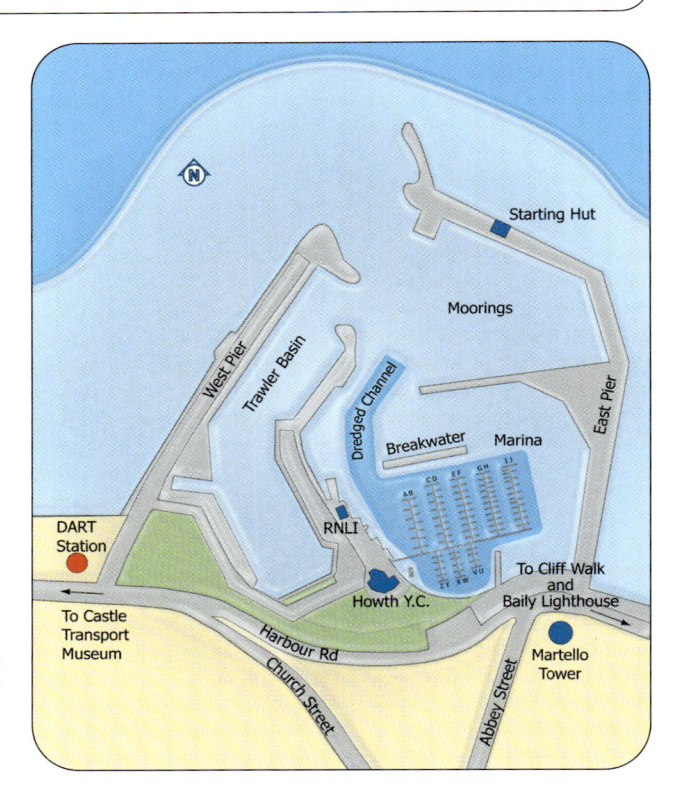

Howth
(see page 113)

Donaghadee

Latitude: 54°38'.50'N; Longitude: 05°31'W;
Charts: Admiralty 3709 2198; Imray C62

Leaving Bangor and Belfast Lough, Donaghadee lies on the eastern coast of the Ards Peninsula. There is a buoyed shipping channel between the mainland and Copeland Island. There is a small marina with approx 50 berths located to the south of the harbour. Care should be taken entering the marina as there is a sill at the entrance and visitors should call the marina on the manager's mobile phone 0802 363382, for instructions re tides and leading marks before attempting it. Entrance should not be made at night or in poor visibility. The office number is 028 9188 2184. There is a toilet and pay phone in the marina. Daily rates are £6 up to 25 feet, £7 26 to 30 feet and one pound for every five extra feet. Weekly rate is £1.50 per foot. There is a 20-ton travelling hoist. Diesel and electricity can be supplied.

Donaghadee is a lifeboat station. It is the nearest port to Scotland and packets used to sail from here to Portpatrick across the sound. A climb up to The Moat, a hill with a monument, and easily recognisable, gives a clear view of the Scottish coast and the Mull of Galloway. Donaghadee is a popular spot for city people to get the sea air. It has banks, a Post Office and supermarket.

Donaghadee has the distinction of having the oldest pub in Ireland, *Grace Neills*, dating back to 1611, (according to the *Guinness Book of Records*). Visitors have included John Keats, Peter the Great and Franz Liszt. More recently ex-Beirut hostage Brian Keenan enjoyed a tipple here. The small entrance belies a large interior with a restaurant serving lunches and dinner. Other pubs worth a visit include Port O'Call and Pier 36 which serves pub food.

Also:

The Moorings Cafe.

Donnelans Hotel, tel: 01247 883569.

Refer to map on page 101.

Portaferry and Strangford Lough

Latitude: 54°22'.80N; Longitude: 05°33'W;
Charts: Admiralty 2156 2159; Imray C62

Strangford Lough is a beautiful inlet, and the largest in the east Irish Sea, separating the Ards Peninsula to the east from the larger part of Co. Down to the west. It is entered between Killard and Ballyquintin Points. It is highly recommended that it be entered on the young flood. The current here can run at seven knots during springs. Once past Audleys Point the tidal currents are considerably slacker. Even the Norsemen were taken by the tidal currents here, hence its name derives from the label they gave it – Strang Fiord, displacing the old Gaelic name – Lough Cuan. There is a marina located on the seafront, with water and power. Overnight rates are £1 per metre. There are showers and laundry facilities available.

There is a regular ferry service to the town of Strangford across the Lough. Portaferry has a range of shops, a bank and a Post Office. The Lough has several hundred islands and is noted for its wildlife. Nearby is Portaferry Castle built by the Savages. Strangford also has a castle, a fine example of a 16th century fortified house which is now a National Monument. About two miles from Strangford is Castle Ward, built in 1765 by the first Lord Bangor. It is somewhat unique in that one facade is Palladian whilst the other is Gothic, a compromise reached after some marital discord between the Wards over styles! The house is now a National Trust property and is open to the public. It has a restaurant.

There are several pubs and eateries:

The Portaferry Hotel has a bar and restaurant, tel: 028 4272 8231.

The Narrows has a restaurant and also offers accommodation, tel: 028 4272 8148.

The Corn Store Restaurant in Castle Street, tel: 01247 29779.

Fiddlers Green Pub in Church Street.

Across in Strangford there is the **Lobster Pot Restaurant** noted for its fish, tel: 01396 881288.

The Lough is well worth exploring for the scenery and wildlife. On Mahee Island in the north west of the Lough are the remains of Nendrum Abbey, founded in the 5th century. It is the best example extant of how early monasteries were planned and laid out. There are no less than ten yachting and sailing clubs in the Lough. Visitors can contact any of the clubs who will advise on moorings, pontoon berths or anchorages and how to make a safe passage.

Portaferry Sailing Club, tel: 01247 728770
Strangford Sailing Club, tel: 01396 86404 or Secretary 01396 613711
Quoile Yacht Club, tel: 01396 612266
Killyleagh Yacht Club, tel: 01396 828250
East Down Yacht Club, tel: 01396 828375
Ringhaddy Cruising Club, tel: 01238 541158
Strangford Lough Yacht Club, tel: 01238 541202
Newtownards Sailing Club, tel: 01247 813426
Kilcubbin Sailing Club, tel: 01247 738422
Down Cruising Club, tel: 01238 541663

Refer to map on page 102.

Ardglass

Latitude: 54°15'.50N; Longitude: 05°36'.30W;
Charts: Admiralty 633 2093; Imray C62

Just north of a famous promontory with its lighthouse, St. John's Point, lies the village and port of Ardglass, in the County of Down. In medieval times it flourished as a major trading centre and hence is surrounded by no less than seven castles, one of which, Ardglass Fort now forms part of the local golf club! The Isle of Man is only 30 miles away and its peaks are visible on a clear day.

Beside the fishing harbour is the 55-berth Phennick Cove marina with 20 visitor berths (tel: 01396 841377). Consult the pilot for entrance directions. There is water and power on the pontoons and toilets, showers and a laundry are available in the marina office block. Diesel supplies can be arranged.

Charges for an overnight stay are £10 up to 30 feet, £12 30 to 35 feet, £14 35 to 40 feet, and pro rata.

The village has a bank, Post Office, grocery shop and taxi services. There is a range of pubs and eateries:

Golden Treasure a Chinese takeaway; **Skippers Rest** for takeaways; **Tidewater Coffee Shop**; **The Moorings Pub** serving bar food; **The Dock Inn**, and **The Lighthouse Bar**.

Carlingford

Latitude: 54°03'N; Longitude: 06°11'.50W;
Charts: Admiralty 44 2800; Imray C62

The town of Carlingford nestles in the picturesque Lough of the same name, between the Cooley Mountains to the south west and the Mournes to the north east. From the entrance between Ballagan and Cranfield Points, which is marked by the Helly Hunter cardinal buoy, the channel is clearly buoyed, and has commercial traffic. Hopefully as the peace process establishes, there will be less naval traffic. Travel against the tide in either direction is difficult and it is best to go in with the flood and out on the ebb. Having passed the Haulbowline Rock with its lighthouse, to the west the port of Greenore comes into view. The red brick building, now in some disrepair was at one time the stately Great Northern Railway Hotel, and a popular holiday destination, with a golf course, still very much played, nearby. The rail connection to Dundalk was closed many years ago, but the port was reopened in the 1960s and is now a busy commercial operation. The marina is located beyond the town – turn to port at the red no. 18 buoy and head back straight for the marina entrance. Call the marina before entering on channel 37, tel: 042 937 3492.

The marina which has 80 berths (30 for visitors) charges £1.20 per metre per day. Weekly charge is £7 per metre. There is 220v power on pontoons and water, diesel and petrol are available. There are phones, showers, laundrette and dryers in the utility building. Work is in progress on doubling the number of berths and it is hoped that they will be ready for the 2001 season.

An imaginary line drawn through the centre of the Lough defines the border between the Republic of Ireland and Northern Ireland. To the north west Warrenpoint and Rostrevor are clearly visible. At the head of the Lough is the town of Newry, once open to shipping via the now closed Newry Ship Canal, which in turn was connected to Lough Neagh by canal.

Carlingford is another town with Viking origins. They developed it as an important trading base. The imposing castle dates from later times – the 12th century, when it was built as a fortification by Hugh de Lacy. It became an important English settlement. It was attacked in 1596, unsuccessfully, by the Gaelic chieftain, Hugh O'Neill. It later became embroiled in fighting between English loyalists and parliamentarians. Carlingford has many interesting buildings reflecting its history. Apart from the imposing St. John's Castle, these include the Tholsel, the Mint and Taffy's Castle, all in the town centre, which is about ten minutes' stroll from the marina.

Dundalk & Carlingford Sailing Club is situated on the far side of the harbour, and welcomes visiting yachtsmen to its bar, and various functions. It runs a busy dinghy and keelboat racing schedule, and visiting yachts can enter. Please contact a club official.

Carlingford Lough has always been renowned for its oysters. The industry went into decline, but has revived in recent times and flourishes today. August sees the Carlingford Oyster Festival, and the succulent bivalve molluscs, with all their potential, can be sampled in the town eateries and pubs (accompanied by Guinness and brown bread of course, how else?!). There are some very pleasant walks clearly signposted through the Cooley Hills. Bikes can be hired in the village which also has a Post Office, bank and grocery shops. There is a heritage centre describing the history of Carlingford, and an adventure sports centre.

Restaurants:

There is a restaurant with bar in the marina building serving lunch and dinner, seafood, Mediterranean cuisine, tel: 042 937 3073.

Jordans Town House Restaurant, in a pleasantly restored building serves local produce, tel: 042 937 3223.

Magee's Bistro, tel: 042 937 3751.

The Terrace Restaurant and Coffee Shop, Newry Street, open 1000 to 1700 daily, serving evening meals on Friday/Saturday, tel: 042 937 3731.

McKevitt's Oyster Catcher Bistro.

Pubs:

P.J.O'Hare's Anchor Pub is popular with visiting yachtsmen. A plate of oysters or queenies (delicious miniature scallops), and pub grub can be eaten in the pleasant courtyard.

Malahide

Latitude: 53°27'.20N; Longitude: 06°09'.50W;
Charts: Admiralty 633 1468; Imray C61 and C62

From Carlingford travelling south, the passage crosses the great expanse of Dundalk Bay. Then Drogheda with its port entrance comes into sight. Clearly visible are the chimneys of Platin, Ireland's largest cement plant. Also visible is the fishing port of Clogher Head. To the east is the Rockabill Lighthouse, a renowned bird sanctuary. Indeed this part of the Irish Sea has a prolific seabird population — guillemots, cormorants, gulls and gannets sweep and dive in constant search for food. Soon the foothills of the Dublin Mountains are clearly visible. To the east is Lambay Island.

The entrance to Malahide Estuary is marked by a red and white fairway buoy, group flashing white. The channel, about one and a half miles in length, has recently been dredged and can accommodate vessels with a draft of 1.2 metres at low tide, in favourable weather conditions. However, because of the continually changing nature of the bar at the estuary entrance, the marina office should be called on channels 37 or 80 for confirmation regarding channel depth, before entering, and for berthing instructions. From the fairway buoy, the channel is marked by a series of port and starboard hand buoys. The marina breakwaters are marked by daymarks and lights. The marina, (tel: 01 845 4255), set in front of elegant apartments, has 350 berths. The charge for visitors is £1.50 per metre per day or £7.70 per metre per week.

A public telephone is located beside the marina office, as are the showers and toilets. Fresh water is available on all pontoons, and electricity 16 or 32 amps at 220/240 volts. Petrol and diesel fuels are available. There are refuse disposal points in the marina area. Boats up to 30 tonnes can be lifted by mobile crane. There is a fully equipped boatyard offering storage. Engine, electrical and hull repairs can be arranged with the marina office.

The village of Malahide lies nine miles north east of Dublin, in the county of Fingal, one of the four administrative regions of greater Dublin. Fingal, which in Gaelic means 'fair stranger', and was how the conquering Vikings described the local inhabitants. Around it, a busy residential suburb of Dublin has developed. But the village still retains the characteristics of a compact country town. The streets are busy with shoppers and others with various errands to do. There is a wide range of shops, banks, boutiques, hairdressers, pubs and restaurants. Anxious punters can be seen crossing from Smyth's pub, which has a dedicated horse racing lounge, to the bookmakers across the road – their next visit hopefully to collect their winnings on that dead certainty!

The village is just a few minutes walk from the marina, and a further 15 minutes will bring you to Malahide Demesne and Castle. To reach the pedestrian entrance, walk along Main Street, go over the railway bridge and the entrance is clearly visible on the left. The Castle dates from the 12th century, and until recent times was still in the possession of the Talbot family. In 1975 the local authority acquired the imposing castle and adjoining grounds from Lord Talbot de Malahide and they are now open to the public. A remarkable discovery was made in the castle in 1949. This was a major collection of the private papers of James Boswell, biographer of Samuel Johnson. The documents including his now infamous diaries were found hidden in a desk and were purchased by and lodged in Yale University. The connection was that the last Lord Malahide but one was a great-great-grandson of Boswell.

The oldest part of the present castle is a keep-like tower. The principle rooms on view are the oak-beamed Great Hall, the Oak Room with its carved panelling, living rooms and bedrooms. It is said that 14 members of the family breakfasted here before departing for the Battle of the Boyne in 1690, and for all it was to be their last meal. A collection of paintings from the National Gallery of Ireland is on view. Outside there are over 200 hectares of parkland with fine walks, including a beautiful botanical garden. Of special interest is the model railway collection of a local enthusiast named Fry, which is one of the largest of its kind in Europe.

Visitors are welcome from 1000 Monday to Saturday and closing times are 1900 April and October, 2000 May and September and 2100 June through to August, Sundays and bank holidays 1100 to 1800. Entrance charges are £3.10 for adults, children £1.65, OAPs £2.60. There is a £8.50 family charge. The gardens are open May to September 1400 to 1630.

Malahide has a great sporting tradition, and this adds to the atmosphere. Close to the centre is Malahide Cricket Club, founded in 1861, where matches are regularly played in the summer. Next door is the ground of St. Sylvester's Gaelic Football and Hurling Club. On the seafront is Malahide Yacht Club which welcomes visiting yachtsmen to its bar and other facilities. Adjacent to it is Malahide Lawn Tennis and Croquet Club. On the nearby Broadmeadow there is a sailing club and school, with an active dinghy and sailboard fleet. Malahide has an 18 hole golf course, and there are numerous other courses within easy reach, including Portmarnock, the famous championship links course which has hosted the Irish Open. There are also fine sandy beaches within walking distance.

There are regular bus and train services to Dublin and other centres. Dublin airport is seven kilometres away. The DART electric train service has been extended to Malahide and services commenced in 2000. This service takes one around Dublin Bay to Bray and on to Greystones in Co. Wicklow, passing through Dublin and Dun Laoghaire.

Souvenirs and gifts: Fragrance Boutique Ltd., located beside the marina office carries a range of quality toiletries, candles and other gift items. The Pottery Shop in Church Street carries a fine range of ceramic and glass giftware.

Restaurants

Malahide boasts a wide range of eateries from Chinese and pizza takeaways to haute cuisine. The following is a small selection:

Cruzzo Restaurant, overlooking the marina. A bit pricey but excellent food and decor. tel: 01 845 0599.

Breakers Restaurant, New Street, wide menu, full licence, lunch Sunday, dinner every day, tel: 01 845 2584.

Giovanni's Pizzeria, Townyard Lane, Solid Italian menu, open lunch/dinner seven days, tel: 01 845 1733.

Les Visages, Marine Court (on sea front), wide menu, dinner Tuesday to Saturday, tel: 01 845 1233.

The Orangerie, Townyard Lane, Fish and Cajun/Mexican dishes, lunch Sunday only, dinner Monday to Saturday, tel: 01 845 1299.

Oscar Taylors, Coast Road, wide menu, lunch Saturday and Sunday, dinner every day, tel: 845 0399.

Sale e Pepe, New Street, open for dinner seven days, wide menu, tel: 01 845 4600.

Sakura, Japanese in New Street, tel: 01 845 6288.

Silk's Restaurant, Main Street, Chinese on the expensive side, dinner only – seven days, tel: 01 845 3331.

Woodhouse Bistro, Coast Road, oak barbecue specialities, tel: 01 845 1988.

Bon Appetit, St. James Terrace, specialities are fish, and game, lunch Monday to Friday, dinner Monday to Saturday, definitely on the expensive side, tel: 01 845 0314.

Smyths Diamond Bar (see under pubs).

For takeaways, try **Beachcomer**, **Bangkok Thai** and **Green Cottage** (Chinese), all in Townyard Lane.

Malahide's three pubs are located close to each other – they are:

Smyths (Diamond Bar), New Street, a lively pub with sports TV viewing and serving a wide range of very reasonably priced dishes all day. Traditional music sessions during the week.

Gibney's Traditional Irish Pub, New Street, Lunch served from 1230, a good range of traditional dishes. Traditional music sessions on Sunday with gigs performed in the afternoon.

Duffy's, Main Street, lunches are served daily, traditional music on Thursday nights, gigs performed regularly in the courtyard during summer.

Hotels:

The Grand, an imposing period building a short distance from the village centre. tel: 01 845 000.

The Stuart, (incorporating Oscar Taylor's restaurant), on the seafront, tel: 01 845 0399.

The village Laundry is in Strand Street; Post Office and Supermarket in Townyard Lane.

Refer to map on page 103.

Howth

Latitude: 53°23'.60N; Longitude: 06°04'W;
Charts: Admiralty 1415 1468; Imray C61 C62

Travelling down the Malahide estuary inlet, Lambay Island is clearly visible four miles to the north east. The island is the private property of the estate of the late Lord Revelstoke. A ferry runs to the island from Rogerstown, but permission must be obtained prior to visiting. There were early Christian settlements on the island, but few traces remain because of the damage done by various invaders including the Vikings. Just partly visible behind trees close to the western side of the island is Lambay Castle designed as a home for the Revelstokes by the famous English architect Sir Edwin Lutyens. On the north coast there are extensive seabird colonies.

Leaving the dredged inlet and turning south, Ireland's Eye island and the Hill of Howth are in view. To the west is Portmarnock Strand and dunes and behind them the famous Portmarnock Golf Club. Ireland's Eye is a bird sanctuary and haunt for day trippers. Masses of gulls, cormorants, terns and guillemots cling to the cliffs.

The deeper water is to the east of Ireland's Eye. The Rowan Rocks buoy is to seaward and the South Rowan Buoy is close to the harbour entrance, which is lit. The harbour is the headquarters of the east coast fishing industry and there is much trawler traffic, to watch out for. Entering the harbour the trawler harbour entrance is to the right, with the marina entrance ahead, approached between red and green markers which are not lit at night, but the street lights supplemented by a spotlight, give light. The welcoming blue and white stripes of the Yacht Club awning, are clearly visible.

Before entering the harbour, the marina should be called on channels 37 or 80. The marina (tel: 01 839 2777) has 300 berths. Charges for visitors are 35p per foot per day, a week's stay is charged at six days. 220v power and water are available on the pontoons. There are visitors' showers and toilets in the clubhouse and a launderette for which tokens should be purchased from the marina office. There is a well stocked chandlery, Dinghy Supplies in Sutton, about three miles away. The club crane will lift up to seven tons. For repairs enquire in the marina office. There is also a drying-out pad.

Howth, like Malahide, is a residential suburb of Dublin, which is about eight miles away. The DART rapid rail system terminates here and can carry you right around Dublin Bay to Dun Laoghaire and then to Greystones via Bray in Co. Wicklow, passing through the centre of Dublin. There are regular bus services to Dublin. The Hill of Howth rising behind the harbour, drops down to a narrow isthmus at Sutton, connecting it to Dublin.

Clearly visible as you enter the harbour to the right are the imposing grounds of Howth Castle and Demesne, still occupied by the St. Lawrence family. The present castle dates from the 16th century. Tradition has it that the family still keep a place at table for the celebrated warrior – Granuaile – of legend (see entry on Clare Island). As a result of finding the castle gate closed to her at dinner time, in revenge she carried off a hostage, only releasing him when promised that the gates would never again be closed to her. There is a public golf course, Deer Park, beside the castle. Also here is the National Transport Museum, housing a collection of trams, buses and other commercial vehicles from days gone by.

Howth Harbour was the terminus for Holyhead packets until 1834, when silting, and the building of Dun Laoghaire led to their relocation there. In 1914, the yacht *Asgard* skippered by Erskine Childers, landed a cargo of guns which were to be used in the 1916 rebellion. Childers, an Englishman and one time clerk of the House of Commons, was attracted to the Irish freedom movement. His book, *The Riddle of the Sands*, is a classic, and was based on his sailing adventures in the Baltic and prophetically described German naval preparations for the First World War. He married Mary Ellen Osgood of Boston and the *Asgard* was her wedding present to him. She accompanied her husband on the gun running trip. Tragically he was to die in the civil war which followed Ireland's independence in 1922. The *Asgard* was subsequently used as a sail training vessel before being retired. Plans are now advanced to restore her. Her successor, the square-rigged brigantine *Asgard II*, is now Ireland's sail trainer.

A public pathway takes one around the eastern side of Howth Head to the Baily Lighthouse from where there is magnificent panoramic view of Dublin Bay, with Dun Laoghaire clearly recognisable by its harbour and church spires, and the great sweep of the Dublin and Wicklow mountains in the background.

Howth has a busy racing and cruising fleet. In winter practically every garden holds a dinghy or larger boat. In summer, sailors can be busily seen doing their domestic duties, mowing the lawn while dressed in sailing gear, before heading for the harbour. The harbour is home to a fleet of 'seventeens'. These are gaff-rigged sloops 17 feet on the waterline, and date from 1898. They are reputed to be the oldest one design keel boat, and were built by John Hilditch in Carrickfergus. They have raced continuously since then and are distinguished by their colourful topsails.

Howth has a good selection of restaurants and pubs and there are four hotels. There are several fish shops on the West Pier where you can buy a wide selection of the freshly landed catch.

Hotels:

The Baily Court, tel: 01 832 2691; **The Deer Park Hotel & Golf Course**, tel: 01 832 2624; **The Howth Lodge**, tel: 01 832 1010; **St. Lawrence**, tel: 01 832 2643.

Restaurants:

Visitors are welcomed to the dining room and bar of Howth Yacht Club. Neat dress is required. Howth is renowned for its fish restaurants, but other tastes are catered for as well:

The **Abbey Tavern**, Abbey Street, **The Abbot Restaurant and Bar**, tel: 01 839 0307, famous for its traditional music sessions.

Adrians Restaurant, Abbey Street, medium priced, wide range of fish and meat, tel: 01 839 1696.

Casa Pasta, 12 Harbour Road, good and reasonably priced pizza spot, also serving salads and seafood, tel: 01 839 3823.

De Gee's, 1 Harbour Road, budget prices, open all day, tel: 01 832 264.

Asian Tandoori, Harbour Road, reasonably priced Indian, tel: 01 832 5966.

El Paso, 10 Harbour Road, good Italian, tel 01 832 3334.

King Sitric, East Pier, renowned for its seafood, but expensive, tel: 01 832 5235.

Lil's Coffee Shop, Harbour Road, tel: 01 839 3045.

Porto Fino's, tel: 01 832 3048.

Aqua @ The Water Club, restaurant located in the old yacht club building on the West Pier with nice views, serves steaks and seafood, tel: 01 832 0690.

Citrus Tree, Harbour Road, for fish and steaks, tel: 01 832 0200.

Dragon Boat, Chinese takeaway, opposite the Dart Station.

Pubs:

The Abbey Tavern, Abbey Street; **The Bloody Stream**, beside the Dart Station, serves food all day; **Cock Tavern**, Church Street; **Evora Lounge**, Church Street; **The Lighthouse Bar**, Church Street; **Waterside Bar**, Harbour Road, with **Wheelhouse Restaurant** attached.

Refer to map on page 104.

Postscript

As so often happens with publications that carry a lot of information, no sooner is it finished than new facts come to life. And questions have been asked as to whether this or that location has been included. As the Irish cruising scene is constantly developing and for the better, the following are some useful additions.

Youghal in east Cork is a delightful old medieval town which contains many places of interest, not least Elizabethan buildings including Sir Walter Raleigh's house and the nearby exquisite Saint Mary's Church which has remained open for worship for 600 years. Youghal was used by director John Huston as the location for his filming of the epic *Moby Dick*. A fisherman may be able to provide a spare mooring. If not, drop anchor near the pier in the vicinity of a line of moorings. Holding is quite good, but a kedge may be needed as the river in spate can run at two and a half knots. Another good reason for visiting Youghal is Aherne's Seafood Restaurant which many well informed palates believe is one of the best in Ireland.

Castletownsend in west Cork is a very pleasant and safe anchorage. The writers Somerville and Ross, who wrote *Some Experiences of an Irish R.M.*, lived in nearby Drishane and were familiar figures in the village. A pleasant stroll up the street brings you to a most welcoming pub and restaurant, Mary Ann's. You will retire to your boat well fed and wined.

Bord Fáilte and The Marine Institute are completing their programme of cruising facilities with pontoons being installed in Skerries, Rush (Rogerstown) and Ballyhack. Skerries is in north Co. Dublin. It has an active sailing fleet and Skerries Sailing Club has a fine clubhouse overlooking the harbour. Planning may result in a marina being constructed here. Skerries is a sizeable town with all the usual facilities and a wide range of pubs and eateries.

Rush Sailing Club has a lovely clubhouse overlooking Rogerstown inlet, also in north Co. Dublin, close to Skerries. Plans are well advanced to install a pontoon. The village of Rush is about a mile walk. All facilities including pubs and eateries can be found here.

Plans are also being finalised to install a pontoon at Ballyhack about five miles north of Hook Head, Waterford Harbour.

When new information comes to hand and changes occur to current entries, they will be entred into the website. Readers who want to keep updated should access the site – **www.cruisingireland.com**

Further information

This is just a selection of useful contacts. Consult the Golden Pages in the Republic and The Yellow Pages in Northern Ireland for further requirements.

Chandleries:

B.J.Marine, Sir John Rogerson's Quay, Dublin 2, tel: 01 671 9300 and Bangor Marina, tel: 01247 271434/271395
C.H.Marine, Glandore, Co. Cork, tel: 028 33267
Dinghy Supplies, Sutton, Dublin, tel: 01 832 2312
Viking Marine, Lower Georges Street, Dun Laoghaire, tel: 01 280 6654
Newmills Car & Marine, Coleraine, tel: 01265 43535
Union Chandlery, 13 Andersen's Quay, Cork, tel: 021 271 643
Windmill Leisure, Windmill Lane, Dublin, (also charts, nautical books), tel: 01 677 2008
Western Marine, Bulloch Harbour, Dalkey, Co. Dublin, tel: 01 280 0321
Kilmore Quay Chandlery, Kilmore Quay, tel: 053 29791

Sailmakers:

Downers, Dun Laoghaire, tel: 01 280 0231
Watson Sails, Malahide, tel: 01 832 6466
Sterling Sails, Bray, Co. Wicklow, tel: 01 286 3401
McWilliams, Crosshaven, Co. Cork, tel: 021 831505
Sketrick Sailmakers Ltd, Kilinchy, Co. Down, tel: 01238 542005

Engine Repairs:

Conor Treacy, Dun Laoghaire, tel: 087 2503087 (mobile), 01 282 7882
Marine Engineering Co, Killiney, Co. Dublin, (Yanmar) tel: 01 282 2023
Greene Marine Ltd., Dublin, (Outboards), tel: 01 492 3061
Noel Filgate, Dublin, (Volvo) tel: 01 455 5392
Killen Marine (Outboards), Dun Laoghaire tel: 01 285 3908

Hulls and Rigging:

Gerry Doyle, Dun Laoghaire (Rigging), tel: 087 677 2779
Kilmacsimon Boatyard, Bandon, tel: 021 775134
Crosshaven Boatyard, tel: 021 831161
Rossbrin Boatyard, Schull, tel: 028 37352
O'Sullivans Marine, Tralee, tel: 066 24524
Csatlepoint Boatyard, Crosshaven, tel: 021 832154

Electronics:

Tony Brown, Dublin, tel: 01 839 5222

Bibliography

Clark, Wallace, **Sailing Round Ireland**, An Anti-clockwise Journey Around Ireland, Batsford, London, 1976 or North West Books, 1990.

de Courcy Ireland, John, **The Sea and the Easter Rising**, Maritime Institute of Ireland, 1966.

Irish Cruising Club, Sailing Directions, South and West Coasts, Irish Cruising Club Publications Ltd., 1999.

Irish Cruising Club, Sailing Directions, East and North Coasts, Irish Cruising Club Publications Ltd., 1999.

Macmillan Nautical Almanac, published annually.

O'Crohan, Tomas, **The Islandman** (An tOileánach).

O'Sullivan, Maurice, **Twenty Years A-Growing** (Fice Bliain ag Fás), Chatto and Windus, London, first published 1933.

Phelan, Andrew, **Ireland from the Sea**, A Clockwise Circumnavigation of Ireland, Wolfhound Press, 1998.

Road Book of Ireland, Automobile Association.

Sayers, Peig, **Peig**, Dublin, 1973.

Severin, Tim, **The Brendan Voyage**, Hutchinson, London, 1978.

Shell Guide to Ireland, first published, London 1962, regularly reprinted.

Index

Adrigole *(page 59)*
Aranmore *(page 86)*
Ardglass *(page 107)*
Arklow *(page 29)*
Ballycastle *(page 92)*
Ballycotton *(page 43)*
Ballyglass *(page 83)*
Baltimore *(page 53)*
Bangor *(page 95)*
Blacksod *(page 82)*
Carlingford *(page 108)*
Carnlough *(page 93)*
Carrigaholt *(page 71)*
Carrickfergus *(page 94)*
Castletownbere *(page 61)*
Clare Island *(page 81)*
Clifden *(page 78)*
Coleraine *(page 91)*
Cork Harbour *(page 44)*
Courtmacsherry *(page 51)*
Crookhaven *(page 57)*
Crosshaven *(page 46)*
Culdaff *(page 90)*
Derrynane *(page 63)*
Dingle *(page 66)*
Donaghadee *(page 105)*
Downings *(page 87)*
Dromquinna *(page 62)*
Dublin *(page 18)*
Dungarvan *(page 41)*
Dun Laoghaire *(page 23)*
East Ferry *(page 44)*
Elly Bay *(page 83)*
Fahy Bay *(page 80)*
Fenit *(page 69)*
Foynes *(page 73)*
Glandore *(page 52)*
Glengarriff *(page 58)*
Helvic *(page 41)*
Howth *(page 113)*

Inisturk *(page 80)*
Kells *(page 66)*
Kilcummin *(page 83)*
Kilkieran *(page 76)*
Kilmore Quay *(page 36)*
Kilronan *(page 74)*
Kilrush *(page 72)*
Kinsale *(page 48)*
Knightstown *(page 66)*
Labasheeda *(page 73)*
Lawrence Cove *(page 59)*
Leenane *(page 80)*
Malahide *(page 109)*
Maumeem *(page 76)*
Moville *(page 91)*
Portaferry/Strangford Lough *(page 106)*
Portmagee *(page 65)*
Portnoo/Iniskeel *(page 86)*
Portsalon *(page 87)*
Rathmullan *(page 89)*
Roundstone *(page 77)*
Schull *(page 56)*
Sherkin Island *(page 55)*
Smerwick *(page 68)*
Sneem *(page 62)*
Bangor *(page 95)*
Struthan *(page 76)*
Teelin *(page 85)*
Ventry *(page 68)*
Waterford *(page 38)*
Wicklow *(page 27)*
Wexford *(page 31)*